HIGH P_T PHYSICS AT HADRON COLLIDERS

This book provides a comprehensive introduction to high transverse momentum reactions at hadron (proton–proton or proton–antiproton) colliders. It begins by introducing the Standard Model of high energy physics and describes the specialized detectors used. It then gives a general treatment of the reactions to be studied and summarizes the state of the art in hadron collider physics, defined by Tevatron results. The experimental program at the detectors being built for the Large Hadron Collider at CERN is described, with details of the search program and the general strategy to find the postulated Higgs particle. Speculations of physics beyond the Standard Model are also discussed. The book includes links to online freeware programs, freeware utilities, and high energy physics library resources. This book is suitable for graduate students and researchers in high energy physics.

DAN GREEN received his Ph.D. from the University of Rochester in 1969. He held a post-doctoral position at Stony Brook from 1969 to 1972 and worked for a time at the Intersecting Storage Rings (ISR) at CERN. His next appointment was as an Assistant Professor at Carnegie Mellon University from 1972 to 1978 during which time he was also Spokesperson of a BNL Baryonium Experiment. He has been a Staff Scientist at Fermilab from 1979 to the present, and has worked in a wide variety of roles on experiments both at Fermilab and elsewhere. He participated in the D0 Experiment as Muon Group Leader from 1982 to 1990 and as B Physics Group Co-Convener from 1990 to 1994. He led the US Compact Muon Solenoid (CMS) Collaboration as Spokesperson and then was Project Manager for the US groups working at the Large Hadron Collider (LHC) at CERN. At Fermilab, he was Physics Department Deputy Head from 1984 to 1986 and Head from 1986 to 1990. From 1993 to the present he has served as the CMS Department Head in the Particle Physics Division.

CAMBRIDGE MONOGRAPHS ON
PARTICLE PHYSICS
NUCLEAR PHYSICS AND COSMOLOGY
22

General Editors: T. Ericson, P. V. Landshoff

HIGH P_T PHYSICS AT HADRON COLLIDERS

DAN GREEN

Fermi National Accelerator Laboratory
Batavia, Illinois

CAMBRIDGE
UNIVERSITY PRESS

CAMBRIDGE UNIVERSITY PRESS
Cambridge, New York, Melbourne, Madrid, Cape Town, Singapore, São Paulo, Delhi

Cambridge University Press
The Edinburgh Building, Cambridge CB2 8RU, UK

Published in the United States of America by Cambridge University Press, New York

www.cambridge.org
Information on this title: www.cambridge.org/9780521120487

First published 2005
This digitally printed version 2009

A catalogue record for this publication is available from the British Library

Library of Congress Cataloguing in Publication data
Green, Dan.
High P_T physics at hadron colliders / Dan Green.
p.   cm. – (Cambridge monographs on particle physics, nuclear physics, and cosmology; 22)
Includes bibliographical references and index.
ISBN 0 521 83509 7
1. Hadron colliders.   2. Proton–proton interactions.   3. Proton–antiproton interactions.
I. Title.   II. Series.
QC787.C59G74   2004
539.7´3723 – dc22   2004045198

ISBN 978-0-521-83509-1 hardback
ISBN 978-0-521-12048-7 paperback

Science is an integral part of culture. It's not this foreign thing, done by an arcane priesthood. It's one of the glories of the human intellectual tradition.

Stephen Jay Gould (1990)

... some of our thinking should reveal the true structure of atoms and the true movements of the stars. Nature, in the form of Man, begins to recognize itself.

Victor Weisskopf (1962)

Contents

Acknowledgments

This text began as a series of lectures given to graduate students in Brazil and then later in expanded form to students at Fermilab. The comments and questions of the students have proved to be invaluable in improving this book. The secretarial work of Ms T. Grozis has made it possible to go smoothly from chaotic lecture notes to a complete and polished text. Finally, the many high energy physicists working on D0 and CMS as collaborators have shared their knowledge and insights countless times.

Acronyms

ALEPH	One of the four experiments at the electron–positron collider (LEP) at CERN.
ATLAS	One of the two general purpose experiments at the Large Hadron Collider (LHC) at CERN.
BaBar	The experiment running at the electron–positron collider (PEP-II) at SLAC.
Belle	The experiment running at the electron–positron collider at KEK, Japan.
Bose–Einstein	The statistics obeyed by integral spin particles.
Breit–Wigner	The mass distribution for resonant (unstable) particles, characterized by a central mass, M_0, and a decay width, Γ.
CDF	One of the two general purpose experiments at the Tevatron proton–antiproton collider at Fermilab (FNAL).
CERN	The European Centre for Nuclear Research located in Geneva, Switzerland.
CKM	The unitary 3×3 matrix which describes the mixing of the strong interaction eigenstates (quarks) appropriate to the weak eigenstates active in decays.
CM	The center of momentum frame, where the total momentum of the system is zero.
CMS	One of the two general purpose experiments at the LHC at CERN.
COMPHEP	A program to do calculations in high energy physics.
CP	The combined operation of charge conjugation (C) and parity inversion (P).
CTEQ	One of the sets of distribution functions of quarks and gluons available in COMPHEP.
D(z)	The fragmentation function describing how a jet fragments into a collection of particles which carry a fraction z of the jet momentum.
DELPHI	One of the four experiments at the electron–positron collider (LEP) at CERN.

Dirac Equation	The equation which describes the motion of spin $\frac{1}{2}$ particles.
D0	One of the two general purpose experiments at the Tevatron proton–antiproton collider at Fermilab (FNAL).
Drell–Yan	A process in which a quark from one hadron annihilates with an antiquark from another hadron to create a boson.
EM	The electromagnetic compartment of a calorimeter.
EW	The electroweak theory which unifies electromagnetic and weak interactions.
Fermi–Dirac	The statistics obeyed by half integral spin particles. Such particles obey the Fermi Exclusion principle where two fermions cannot be in the same quantum state.
Fermilab (FNAL)	A national laboratory in the USA with a proton accelerator, located in Illinois.
Feynman diagram	A diagrammatic representation of a fundamental interaction as it occurs in space and time.
Γ	The decay width for a specific process. The inverse of the decay width is the lifetime for that process.
G	The effective coupling constant for a four fermion weak interaction.
GUT	A Grand Unified Theory, which proposes to unify the electroweak and strong interaction theories at a high mass scale.
HAD	The hadronic compartment of a calorimeter.
Higgs	The proposed particle in the Standard Model which has spin zero and whose field takes on a non-zero expectation value in the vacuum.
Jet	A collection of "stable" particles which is the result of the fragmentation of a quark or gluon. The jet has a parent quark or gluon.
KEK	The Japanese center for high energy physics located in Tskuba, Japan.
Klein–Gordon	The equation which describes a relativistic particle that has no intrinsic spin.
Λ	A parameter that is defined to "cutoff" the mass scale for a theory, which is badly behaved outside that mass range.
L3	One of the four experiments at the electron–positron collider (LEP) at CERN.
LC	Linear Collider. This is a proposed electron–positron collider which will have sufficient energy to directly form the Higgs boson.
LEP	The Large Electron Positron collider which operated at CERN prior to the LHC construction.

LHC	The Large Hadron Collider is a proton–proton collider with a total CM energy of 14 TeV, which will begin operations at CERN in 2007.
Λ_{QCD}	The cutoff energy in QCD below which the strong interactions become very strong. The strong interactions approach zero strength at very high energy scales.
LSP	Lightest supersymmetric particle. In SUSY models the LSP is usually assumed to be absolutely stable. Therefore, the LSP is a candidate for the dark matter.
Luminosity	The quantity which when multiplied by the cross section gives the reaction rate for a particular process.
Monte Carlo	A numerical technique to simulate processes by choosing various quantities from specified distribution functions.
MMSM	Minimal SUSY model. The SUSY hypothesis has many possible realizations. The minimal model allows for specific predictions by making assumptions having minimal extensions to the SM.
MRS	One of the sets of distribution functions of quarks and gluons available in COMPHEP.
OPAL	One of the four experiments at the electron–positron collider (LEP) at CERN.
Pseudorapidity	A variable which approaches rapidity for particles with mass less than transverse momentum and which is a function only of polar angle.
P_T	Transverse momentum. The transverse direction is perpendicular to the incident hadrons in a *p–p* collision. Therefore, a large value of this parameter indicates a violent collision probing small distance scales.
PYTHIA	A Monte Carlo program which has a model for the fragmentation of quarks and gluons and also models the "underlying event" caused by the fragments of the fractured hadrons.
QED	Quantum electrodynamics. The relativistic theory of the interaction of photons with fundamental, point like fermions.
QCD	Quantum chromodynamics. The relativistic theory of the interaction of colored gluons with fundamental point like colored quarks.
R, G, B	The color labels, red, green, and blue. The assignments are arbitrary and simply label the three distinct color charges contained in SU(3).
Rapidity	A variable used in high energy physics because it is additive under "boost" or Lorentz transformation. It is the relativistic

	analog of velocity. One particle phase space implies a uniform rapidity distribution.
SLAC	The Stanford Linear Accelerator Center located in California and operating an electron–positron collider.
SLD	The detector operating at the SLAC Linear Collider (SLC) which studied the properties of the Z by resonant formation.
SM	Standard Model. The model of high energy physics which describes fermions, quarks and leptons, interacting by electroweak interactions and strong interactions. Gravity is not included in the SM.
SUGRA	A simplified SUSY model which has only five free parameters, thus giving predictive power.
SUSY	Supersymmetry. The postulated symmetry relates fermions to bosons. Therefore, each particle in the SM has a partner, all as yet undiscovered.
Tevatron	The proton–antiproton collider operated at Fermilab at a CM energy of 2 TeV.
UA1, UA2	The underground area experiments using the proton–antiproton collider operated at CERN at 0.27 TeV.
V-A	The coupling of weak currents to fermions is by way of vector and axial vector interactions. Parity and charge conjugation violation are an immediate consequence of the V-A form.
Weinberg angle	The angle that specifies the unitary rotation of the basic neutral gauge bosons into the physically realized Z and photon.
WW fusion	The process whereby a quark emits a W from both initial state hadrons, thereby initiating W–W scattering or fusion into a variety of final states.
Yukawa interaction	A linear interaction between two fermions and a boson. If the boson has a mass, the range of the interaction is limited in space.
Z* (W*)	"Off shell" gauge bosons which are out on the allowed, but improbable, "tail" of the Breit–Wigner mass distribution.

Introduction

Overview

The Standard Model (SM) of high energy physics has been one of the great syntheses of the human intellect. It began about a century ago with the discovery of the electron, which was the first fundamental point like particle to be discovered. In the last decade, the elusive top quark and the τ neutrino have been observed. The sole remaining undiscovered particle predicted by the SM is the Higgs particle, whose vacuum field is believed to give mass to all the particles in the Universe. This text concentrates on the search for the Higgs particle at proton–(anti)proton colliders, those accelerators that collide protons and (anti)protons head on. Indeed, there are complementary efforts at electron–positron colliders, but they are outside the scope of this book.

In outline, Chapter 1 concerns itself with a summary of the Standard Model (SM), giving the particles comprising the SM and their interactions. Mathematical detail is relegated to Appendix A. Chapter 1 closes with twelve questions which are unanswered in the SM but which appear to be of fundamental importance. The next four chapters are concerned with the two initial questions that refer to electroweak symmetry breaking and the Higgs boson.

In Chapter 2 we explore a "generic" general purpose detector, which is representative of those in use at proton–(anti)proton colliders. Specifically, we examine the extent to which the SM particles introduced in Chapter 1 can be cleanly identified and measured. The accuracy with which the vector momentum and position of a SM particle can be measured is very important, as it will influence search strategies for the Higgs.

Chapter 3 is concerned with the specific issue of particle production at a proton–(anti)proton collider. The relevant formulae are given that will enable the student to estimate reaction rates for any process. In addition, the COMPHEP program can then be used to refine the initial estimates. However, students are strongly encouraged to start with the "back of the envelope" estimate before invoking COMPHEP or any other Monte Carlo program. COMPHEP is explained in Appendix B and is readily available to the student, as discussed in the section on tools below. Kinematic details are placed in Appendix C.

Chapter 4 follows up with a discussion of how recent data taken at colliders inform on the predictions of the SM. This section is a snapshot of the present state of the art in the physics of high transverse momentum phenomena as explored at proton–(anti)proton colliders.

In Chapter 5 we start to venture beyond the bounds of current data. This entire chapter is devoted to the upcoming search for the elusive Higgs boson. Much of the presentation concerns itself with the Large Hadron Collider (LHC) at the European Centre for Nuclear Research (CERN) because this facility, slated to become operational in 2007, was specifically designed to search for, and discover, the Higgs scalar (spin zero). Nevertheless, we will see that the search may be long and arduous.

Finally, in the last chapter, we return to the remaining ten fundamental questions raised in the first chapter. Some hint of theories beyond the SM and their consequences is given. In particular, the possibility is discussed that a new symmetry of Nature, a supersymmetry (SUSY) relating space-time and particle spin, might be discovered in the near future.

Scope

The mathematical complexity used here is no more than calculus. However, the concepts used require a good knowledge of quantum mechanics, special relativity and some acquaintance with field theory. Knowledge of Feynman diagrams will be essential, in part because examples of Feynman diagrams are given in the text and also because COMPHEP supplies diagrams for any process that is specified. The intended audience is then advanced graduate students or research workers in particle physics. Full theoretical rigor has, however, been sacrificed in an attempt to reach as wide and as young a group of students as possible.

Units

In this text, we will use units that are common in high energy physics. The Planck constant, $\hbar$, has the dimensions of momentum (P) times length (x) or energy (E) times time (t). (Recall the Heisenberg uncertainty relations $\Delta x \Delta P_x \geq \hbar$, $\Delta E \Delta t \geq \hbar$.) Thus $\hbar c$ has the dimension energy times length and numerically is 0.2 GeV fm. The energy unit used herein is the electron Volt (eV), the energy gained by an electron in dropping through a potential of 1 volt, and 1 GeV $= 10^9$ eV. The unit of length that is most commonly used is 1 fm $= 10^{-13}$ cm, which is the approximate size of a proton.

Other quantities with energy units are proportional to mass (m), mc^2, and momentum, cP. We adopt units with $\hbar = c = 1$. In these units, mass is given in GeV, as is momentum. For example, the proton mass is 0.938 GeV. Length, x, and ct have the dimensions of inverse energy, using $\hbar c$. We will use the notation [] to indicate the dimensions of a quantity. It should be easy for the reader to restore units by replacing P with cP, m with mc^2 and so forth.

Recall that coupling constants indicate the strength of the interaction and characterize a particular force. For example, electromagnetism has a coupling constant which is the electron charge, e, and a "fine structure" constant $\alpha = e^2 / 4\pi \hbar c$ that is dimensionless. The electromagnetic potential energy is $U(r) = eV(r) = e^2/r$ and $V(r)$ is the electromagnetic potential. The dimensions of e^2 are then energy times length, $[e^2] = [U(r)r]$, the same as those of $\hbar c$. Thus, in the units we adopt, $\hbar = c = 1$, e is also dimensionless. With $\alpha \sim 1/137$, we find $e \sim 0.303$. Coupling constants for the two other forces, the strong and the weak, will be indicated by g_i, and the corresponding fine structure constants by α_i, with $i = s$, W.

The units for cross section, σ, which we will use in this text are barns (b) (1 b = 10^{-24} cm^2). Note that $(\hbar c)^2 = 0.4$ GeV^2 mb where 1 mb = 10^{-27} cm^2. The units used in COMPHEP are pb = 10^{-12} b for cross section and GeV for energy units. As an example, at a center of mass, CM, energy, $\sqrt{s}$, of 1 TeV = 1000 GeV, in the absence of dynamics and coupling constants, a cross section scale of $\sigma \sim 1/s \sim 400$ pb is expected simply by dimensional arguments.

Tools

In this book we have extensively used a single computational tool, COMPHEP, both in the examples given in the text proper, and in the exercises. The aim was to expand the range of the text from a slightly formal academic presentation to a more interactive mode for the student, giving "hands on" experience. The plan was that the student would work the examples given in the text and the exercises and then be fully enabled to do problems on his or her own. COMPHEP runs on the Windows$^{\circledR}$ platform, which was why it was chosen since the aim was to provide maximum applicability of the tool. A LINUX version is also available for students using that operating system.

The COMPHEP program is freeware. We have taken the approach in the text of first working through the algebra. That way, the reader can make a "back of the envelope" calculation of the desired quantity. Then he/she can use COMPHEP for a more detailed examination of the question. The use and description of COMPHEP is explained in detail in Appendix B, where a fully worked out example is given. A web address where the executable code (zipped) and a users' manual are available is also shown in Appendix B. The author also posts these items at uscms.fnal.gov/uscms/dgreen. Freeware to unzip files can be found at www.winzip.com/ and www.pkware.com/.

A word now about the availability of references. The use of internet archives is rather advanced in high energy physics, and we have attempted to make them easily available to the reader. The reader with Web access will have immediate access to the research literature. One of the best places to search is at the Los Alamos site, xxx.lanl.gov. Looking under "Physics" to "High Energy Physics – Experiment" (hep-ex) allows us to search on author, explore new preprints, recent preprints, or abstracts, or search in topics of our choice using the "find" feature. Many of the references cited at the end of each chapter of the text refer to this site, making the papers then directly available to the student.

Free programs to read the file formats used in archiving the research papers, .ps and .pdf, are also available on the web. For example, "pdf" files are read by freeware available at www.adobe.com/. "Postscript," or .ps, files can be read using the download from www.wisc.edu/~ghost/.

Another useful site, which is extensively quoted in the references, is the Fermilab preprint library, fnalpubs.fnal.gov, where the Fermilab references can be downloaded.

Clicking on "preprints" and then on "search" you can look for authors and/or titles and then download the full paper. An exercise is included in Chapter 1 that gives the student practice in accessing the literature.

A compendium of data in high energy physics can be found at the Particle Data Group site, pdg.lbl.gov. Finally, available at www.AnnualReviews.org are full review articles, which allow the student to explore some of the longer articles given in the references.

Our aim is obviously to make the information more immediate for the reader. In addition, some of the references given at the end of the six sections of this text are actual books. They, in turn, are rich sources of knowledge within themselves and sources of additional primary references.

1

The Standard Model and electroweak symmetry breaking

It is better to know some of the questions than all of the answers.

James Thurber

No theory is good except on condition that one use it to go on beyond.

André Gide

1.1 The energy frontier

High energy physics concerns itself with the study of fundamental particles and the interactions among them. Progress in high energy physics in the past was often due to an increase in the available energy for the production of massive particles. Since colliding two objects head-on maximizes the total center of mass (CM) energy and hence the energy available for new particle production, we specialize in this text in colliders as opposed to beams striking "fixed" targets at rest in the laboratory. We are also interested in high mass phenomena, which typically lead to particles at high momentum transverse to the axis of the colliding particles. Thus, we concentrate on the very rare high transverse momentum/energy (P_T or E_T) reactions at colliders.

In Fig. 1.1 we show the available energy for making particles as a function of the year when an accelerator began operation for the last thirty years of high energy physics research. Note the exponential increase in energy as a function of time. That increase has driven the rapid progress in the field. There are two distinct curves, one for proton–(anti)proton colliders and one for electron–positron colliders. In this text we must, in the interests of brevity, confine ourselves to the former. Also in Fig. 1.1 we show the masses of the quarks and force carriers (gauge bosons) with masses >0.1 GeV discovered over the recent past, and a schematic representation of the range of possible Higgs boson masses.

Note particularly that there has been a steady stream of discoveries of new fundamental particles of ever-heavier mass. This progression culminated recently in the discovery of the top quark, of mass 175 GeV, at Fermilab in 1996. Looking into the future, the Large Hadron Collider (LHC) at the European Centre for Nuclear Research (CERN), has been designed to fully cover the mass range where the Higgs boson is thought to exist. Note that the constituent CM energy of Fig. 1.1 is less than the proton–(anti)proton CM energy for reasons we will explain in Chapter 4 and Appendix C.

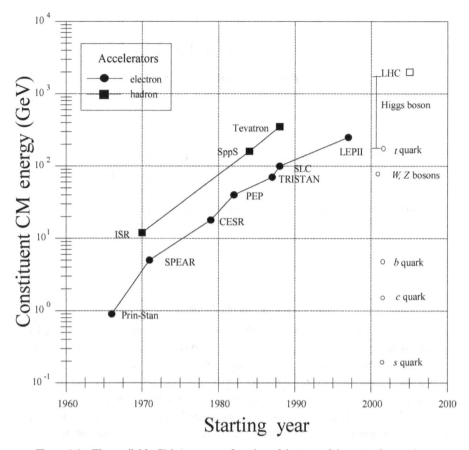

Figure 1.1   The available CM energy as a function of the year of the start of operations of an accelerator. Note the two parallel exponential trajectories for hadron, or proton–(anti)proton, and lepton, or electron–positron, colliders. The masses of the quarks and gauge bosons are also shown.

Therefore, it is timely to first briefly summarize the great accomplishment of particle physics, which is the Standard Model (SM) of fundamental processes. Following that, we can look ahead to the search for the Higgs boson, which will be made possible by yet another advance in the energy frontier.

1.2 The particles of the Standard Model

In the last century, relativity and quantum mechanics were combined together to create quantum field theory. This has led to many insights. For example, each particle is required to have an antiparticle. The first antiparticle to be discovered was the positron, the partner of the electron. In what follows we implicitly assume that each particle has an antiparticle partner, indicated as, for example, $\bar{q}$ being the antiquark partner of the quark, q.

The other great advance of the last century, General Relativity, has resisted inclusion within the SM framework. Thus, at present the SM of high energy physics does not contain gravity as a fundamental quantum theory. Clearly, then, the SM is not a complete theory of Nature.

All three of the forces found in the Standard Model are renormalizable, meaning that calculations in quantum field theory give finite results, while gravity does not. This can be anticipated by observing that classically the "fine structure" constant for gravity, α_G, increases as the square of the mass scale. This follows from noting that the gravitational potential energy, $U_G(r) = G_N M^2/r$, depends on mass, in comparison to the electrical energy, $U_{EM}(r) = e^2/r$. The quantity G_N is Newton's gravitational constant. The fine structure constants of the forces appearing in the SM, such as electromagnetism, where $\alpha = e^2/4\pi\hbar c \sim 1/137$, are dimensionless and mass independent. The gravitational analogue, $\alpha_G = G_N M^2/4\pi\hbar c$, is not.

The SM particles consist of the spin $\frac{1}{2}$ (i.e. $J =$ intrinsic angular momentum $= \hbar/2$) fermions (obeying Fermi–Dirac statistics) which are the matter particles and the spin 1 bosons (obeying Bose–Einstein statistics), which are the force carriers that communicate the forces between the fermions. A listing of these particles as understood today is given in Fig. 1.2. The strongly interacting fermions are called quarks. They are organized as "doublets" with electric charge Q/e, in units of the electron charge, e, of $2/3$ and $-1/3$. The fermions with only electroweak interactions are called leptons. The uncharged leptons, which then have only weak interactions, are called neutrinos.

Let us first consider the fermions, beginning with the quarks. The lightest quarks, the up (u) and down (d) quarks, combine to form familiar bound states like the neutron (udd) and proton (uud) which are held together by the strong force. The quarks are believed to be bound permanently in the proton, say, by the strong force. Ordinary matter is made up of the u and d quarks, which comprise the first "generation." The heavier quarks have larger masses, see Fig. 1.1, but otherwise respond universally to the strong force. They are distinguished by a "flavor" quantum number, which is the weak interaction analog of "electric charge." These heavier quarks comprise the second and third generations. Particles containing strange quarks were seen in cosmic ray events in the 1950s. The charm quark (c) was discovered in 1974, the bottom (b) quark in 1977 and the top quark (t) in 1996.

The leptons are the fermions that do not have the strong "charge" (called "color") as the quarks do. The lightest charged lepton, the electron, has been known for more than a century. It was discovered by J. J. Thompson in 1896. The leptons in Fig. 1.2 are negatively charged; the electron is defined to be a particle, the positron an antiparticle. The other charged leptons appear to be simply heavier "copies" of the electron, all having the same interactions. ("Who ordered that?", as I. I. Rabi was heard to say when the muon was discovered.) The charged lepton masses for e, μ, and τ are 0.5 MeV, 0.105 GeV, and 1.78 GeV respectively. As with the quarks, the leptons comprise pairs of three recurring generations. The tau lepton was discovered in 1975.

The uncharged leptons are called neutrinos and they interact only weakly, having neither "color" nor electric charge. The radioactive "beta decay" of nuclei has also been

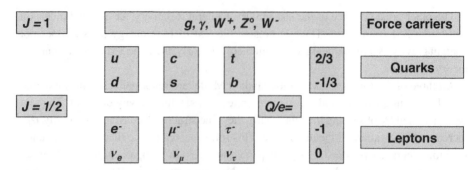

Figure 1.2   The fundamental particles of the SM. The force carriers are spin 1 bosons. The particles of matter are spin $\frac{1}{2}$ fermions. The spin is indicated by the value of J, while Q/e is the electric charge in units of e.

known for a century. These decays were the first evidence for the existence of a "weak force" which caused the conversion of a proton into a neutron and a positron. Neutrinos were hypothesized to be emitted in these weak decays, $p \rightarrow n + e^+ + \nu_e$, but their very low interaction probability made their direct experimental detection a fairly recent phenomenon. The electron neutrino was observed in 1953 near a reactor which supplied a copious source of neutrinos. The tau neutrino was recently seen at Fermilab in 2000. The masses of the neutrinos are measured to be very small and for our present purposes are assigned a zero mass. We return to this topic in Chapter 6. Neutrinos also have "flavor" and come in three distinct varieties, paired to the charged leptons, as seen in Fig. 1.2.

We now turn to the force carriers of the SM. The forces are carried by vector ($J = 1$) bosons, ($[J] = [\hbar] = 1$). The massless quantum of the electromagnetic field, the photon, has also been known as a fundamental particle for almost a century following the explanation of the photoelectric effect by Einstein in 1905. The strong force is carried by massless "gluons" (g) that carry "color," the strong force analog of the charge of electromagnetism. The electromagnetic force is carried by the neutral photon (γ), and the weak force by the W^+, Z^o and W^-, which carry "flavor," the weak force analog of electric charge.

The strong force is needed to explain why the Rutherford nucleus is bound, since electrostatic repulsion of the protons in the nucleus would otherwise break it apart. Gluons were first seen experimentally in the 1970s when they were detected as radiation in electron–positron collisions yielding a quark–antiquark pair and a gluon in the final state, $e^+ + e^- \rightarrow q + \bar{q} + g$. There are eight gluons, each with a distinct color combination.

The weak force is responsible for radioactive decay, where the nuclear charge changes accompanied by the emission of an electron and an antineutrino, $n \rightarrow p + e^- + \bar{\nu}_e$. The force was initially thought to be weak because the decay rates for this "beta decay" were very slow with respect to those of electromagnetic decays. A complete understanding of the dynamics of weak interactions awaited the discovery of the W and Z bosons at CERN in 1983. The masses of the W and Z are $\sim$80 and 91 GeV respectively. The way the W and Z obtain this mass is called "electroweak symmetry breaking" and is thought to be brought about by the "Higgs mechanism." The search for the Higgs is the central theme of this book.

The electromagnetic quantum, or photon, couples to charge, the gluons couple to "color" charge and the W and Z bosons couple to weak "flavor" charge. Gluons are "flavorblind," so all quarks dynamically interact with gluons with the same forces up to the purely kinematic effects of their different masses. The "flavor" quantum number is therefore conserved in the strong interactions, which means that heavy flavors must be strongly produced in particle–antiparticle pairs. The weak interactions are "colorblind" so that the three colors of quark all have the same weak interactions.

At this time the only undiscovered particle known to be required in the SM is the Higgs boson. This is hypothesized to be a fundamental spin 0 field quantum, one that does not appear in Fig. 1.2. It was invented to be responsible for giving mass not only to the W and Z bosons but also to the fermions of the SM. This brief introduction completes the inventory of the "periodic table" of the SM of high energy physics, indicating all the known fundamental particles.

1.3 Gauge boson coupling to fermions

So far, the SM particles have been given more or less as static objects lodged in the high energy physics "table of the elements" displayed in Fig. 1.2. To bring them to life we need to explore their dynamics. There is a great organizing principle for interactions in the SM called "gauge symmetry." We will not proceed from this first principle, but will take a short cut and move ahead by exploiting an analogy to the very successful field theory of electromagnetism. Therefore, just as in electromagnetism, we expect massless vector boson quanta universally coupled to the fermions.

Another force that is very familiar to us is gravity. General relativity asserts that physics is the same in any general coordinate system. That in turn requires the existence of a spin 2 massless "graviton" quantum coupled universally to mass with Newton's coupling constant, G_N.

Therefore we again, by analogy, might expect the weak and strong forces to have massless vector quanta with universal coupling. What, precisely, specifies the interaction of the bosons with the fermions? We again appeal to electromagnetism. In classical mechanics in the Hamiltonian formulation, the student has presumably seen that the free particle Hamiltonian is converted to one describing fermions interacting with photons by the replacement of the momentum $\vec{P}$ by $\vec{P} - e\vec{A}$, where $\vec{A}$ is the vector potential of the electromagnetic field.

The formulation of the electromagnetic interactions in non-relativistic quantum mechanics is the same, where $P \to i\hbar\partial$ is the classical to quantum replacement, as should also be familiar to the student. Therefore, to describe electromagnetic interactions the ordinary derivative ∂_μ is replaced by the "covariant" derivative D_μ in the free particle Lagrangian. The Greek subscript μ is used for indices running from 1 to 4, the standard notation for relativistic equations.

$$\partial_\mu \rightarrow D_\mu = \partial_\mu - ieA_\mu. \tag{1.1}$$

The photon then couples to all the charged pairs that exist in the SM. The fundamental interaction vertices, which appear in the Feynman diagrams, contain two fermions and a boson with a coupling strength of e in the reaction amplitude. The strength of the coupling is universal and is αQ^2 in the reaction rate, where the charge, Q, of the quark or lepton was shown in Fig. 1.2. Schematically, the photon coupling to quarks and leptons is shown in Eq. (1.2).

$$\gamma q\bar{q}, \gamma \ell^+\ell^-. \tag{1.2}$$

The strong interactions have a very similar coupling scheme for the massless colored gluons interacting with the colored quarks. The strong coupling constant is g_s, with strong fine structure constant α_s, which has a value ~ 0.1, about 14 times larger than the electromagnetic fine structure constant, α, as befits the strong force. The Feynman vertices for the strong force have the gluon, g, coupling to quark–antiquark pairs. The amplitude is proportional to g_s. The gluon, g, coupling to quarks, q, is schematically indicated below:

$$g q\bar{q}. \tag{1.3}$$

For the weak force, there are charge changing, beta decay, interactions caused by the charged W bosons and neutral weak interactions mediated by the neutral Z. In fact, we now realize that the "weak" interactions are not intrinsically weak. They are, indeed, unified with electromagnetism and have roughly the same strength. Therefore, we speak of the unified "electroweak" force. The fine structure constant for the weak force is $\alpha_W \sim 1/30$ and the unification of the forces is embodied in the relationship $e = g_W \sin\theta_W$, $\alpha_W = g_W^2/4\pi$, defined by the Weinberg angle, θ_W, a quantity whose magnitude is of order one. The value of the Weinberg angle is not predicted by the SM and must be measured experimentally. It has the observed value, $\sin\theta_W = 0.475$.

The interaction vertices for the charged and neutral weak interactions are:

$$W^- q\bar{q}', W^- \ell^+ v_\ell, Z q\bar{q}, Z\ell^+\ell^-, Z v_\ell \bar{v}_\ell. \tag{1.4}$$

In general, the W can couple to all charged quark pairs, $q\bar{q}'$. However, the most probable pairs are measured to be $W^- u\bar{d}$, $W^- c\bar{s}$, and $W^- t\bar{b}$. The coupling of the Z is to flavorless pairs of quarks and leptons, $\ell = e, \mu, \tau, v_e, v_\mu, v_\tau$.

In non-relativistic quantum mechanics the reaction matrix element is the interaction potential, $V(r)$, bracketed by free plane wave initial and final states in the Born approximation. The amplitude is thus the Fourier transform of the interaction potential, $V(q)$. We appeal again to the case of electromagnetism because it should already be familiar to the student. The Coulomb potential, $V(r) \sim 1/r$, and the photon "propagator," $V(q) \sim 1/q^2$, for the massless photon should be familiar, where q is the magnitude of the difference of vector momentum between the initial and final fermion states, the "momentum transfer."

For example, Rutherford scattering has a reaction amplitude $\sim V(q)$, or a cross section with characteristic behavior $\sim 1/q^4$.

For a particle of mass M, the Fourier transform again gives the transition matrix element, A, in momentum transfer, or q, space. The range λ (the Compton wavelength) is $\sim 1/M$ so that heavy quanta are localized in space and have small reaction rates, $\Gamma \sim |A|^2 \sim V(q)^2 \sim 1/M^4$, *for* $q \ll M$.

$$V(r) \sim e^{-Mr}/r,\ V(q) \sim 1/(q^2 + M^2).\qquad(1.5)$$

The W boson must have a large mass in order to make the interaction appear to be weak and short ranged. The Yukawa form of the interaction potential of a massive vector boson of mass $M \sim 1/\lambda$ is $V(r) \sim [\exp(-r/\lambda)/r]$, which is weak at large r due to the exponential factor but is roughly Coulomb like, $V(r) \sim 1/r$ for $r \ll \lambda$. The effective range of the force is $\lambda \sim 0.0025$ fm for an 80 GeV W mass. At an energy scale of 1 GeV, the exponential reduction factor is about 10^{-36}, which explains why nuclear beta decay appears to be weak (long lifetimes, small decay rates). It required the advent of accelerators of sufficient energy, comparable to the W mass, for us to realize that electromagnetism and weak interactions were aspects of the same force, exhibiting the same intrinsic strength.

1.4 Gauge boson self-couplings

We assume in what follows that all ordinary derivatives that appear in the free particle Lagrangian are to be replaced by "covariant derivatives" which contain the coupling constants and the fields of the gauge bosons. This procedure is done in analogy to electromagnetism. There is an immediate implication of the gauge prescription for replacement of an ordinary derivative by a covariant derivative in the Lagrangian. The term in the Lagrangian representing the free particle kinetic energy for a boson field is quadratic in the field and the derivative. This follows from the relativistic relationship of energy, momentum, and mass (see Appendix C); $E = \sqrt{P^2 + M^2}$, $P_\mu \cdot P^\mu = M^2$, and the quantum mechanical operator replacement, $P \to i\partial$ which then yields the Klein–Gordon Lagrangian density, ℓ, appropriate to bosons, $\ell = (\partial\overline{\phi})^*\partial\phi - M^2\overline{\phi}\phi$, which has a "kinetic energy" term and a mass term.

To describe quantum fields we will use ψ for fermion ($J = \frac{1}{2}$) fields, ϕ for scalar ($J = 0$) fields, and φ for vector ($J = 1$) gauge fields in this text. For masses, m is used for fermions and M for bosons.

Therefore, for a vector gauge field, φ, with coupling constant g, the free kinetic energy under gauge replacement, $D = \partial - ig\varphi$, $\varphi = W, Z, g$, yields trilinear and quartic couplings, as shown schematically in Eq. (1.6). For the familiar case of electromagnetism, since the photon has no electric charge, these self-couplings are absent. However, for the gluons, which carry color charge, and the weak bosons, which possess flavor charge, these couplings are predicted in the SM and lead to measurable cross sections due to the new interaction terms in the Lagrangian density:

$$(\partial \overline{\varphi})^*(\partial \varphi) \rightarrow (D\overline{\varphi})^*(D\varphi)$$

$$\ell_I \sim g(\partial \varphi)\overline{\varphi}\varphi, \quad g^2 \overline{\varphi}\varphi\overline{\varphi}\varphi. \tag{1.6}$$

Although self-coupling is absent for photons, this situation is not completely novel in classical physics. An example, which should be familiar to the student, appears in general relativity. The binding energy of gravity must have mass by the equivalence principle, since all energy is equivalent to mass. Thus the gravitational field itself gravitates; it has gravitational "charge" equal to its mass. In general relativity this results in the classical Einstein non-linear field equations.

In the case of W, Z, and g, by analogy with gravity, the fact that they carry "charges" means that they will self-couple. These interactions between the gauge bosons exist, even in the absence of matter (fermions). They are indicated schematically in Eq. (1.7), which represents the fundamental vertices that can occur in a Feynman diagram:

$$gg g, \; g g g g,$$
$$W^+ W^- \gamma, \; W^+ W^- Z,$$
$$W^+ W^- \gamma\gamma, \; W^+ W^- \gamma Z, \; W^+ W^- ZZ, \; W^+ W^- W^+ W^-. \tag{1.7}$$

1.5 COMPHEP evaluation of boson self-couplings

We have just completed a whirlwind summary of the SM. It is at this point that we can slow down and start to join current research in high energy physics. In this text we will use the computer code COMPHEP, developed at Moscow State University, to get numerical results for SM processes. The code can be used to evaluate both decays and two body collisions into any number of final states. It is available in a Win98® or higher version that will run on any personal computer using this most common of operating systems, Windows. The student is very strongly encouraged to download the code now, read the users' manual, do the exercises of Appendix B, and from now on follow and reproduce the examples shown in the text. The student can, in this way, get a "hands on" experience of up to date research in high energy physics and enhance the utility of the text per se.

There is recent strong experimental evidence for the existence of triple gauge boson couplings from electron–positron collider experiments. In the particular case of WW pair production in electron–positron annihilations, the Feynman diagrams (available in COMPHEP) are shown in Fig. 1.3 for the process $e^+ + e^- \rightarrow W^+ + W^-$. Triple $W^+ W^- \gamma$ and $W^+ W^- Z$ couplings, of the photon and the Z to W pairs, are involved.

The cross section given by COMPHEP is shown in Fig. 1.4 as a function of the available CM energy. Note the rise from threshold at about twice the W mass. Since the W is unstable under weak decay, it has a finite lifetime τ and hence a finite mass width $\Gamma \sim \hbar/\tau$. This width makes for a slow rise of the cross section from the threshold for W pair production, smearing the rise over a CM mass range of about $2M_W \pm \Gamma$.

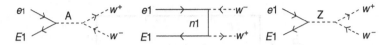

Figure 1.3   Diagrams for electron–positron annihilation into *W* pairs in COMPHEP.

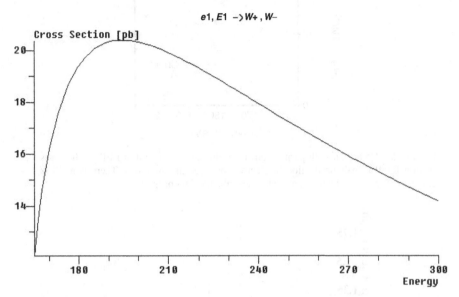

Figure 1.4   Monte Carlo program results for the *WW* cross section as a function of CM energy in electron–positron annihilations.

Experimental data from the CERN Large Electron–Positron collider (LEP) are shown in Fig. 1.5. The agreement with the COMPHEP prediction (Fig. 1.4) is quite good, indicating the experimental confirmation of the predicted triple gauge boson couplings. We also see in Fig. 1.5 that the cross section for simple neutrino exchange (Fig. 1.3) is larger than the full SM cross section. Therefore, a quantum mechanical destructive interference between amplitudes is required to describe the experimental data. The COMPHEP tool has thus let us quickly get up to speed in examining current results in high energy physics.

There are also LEP data on *Z* pair production, which are displayed in Fig. 1.6. Note that the cross section level is about an order of magnitude lower than that for *WW* pair production. Exploring this data is interesting because the SM predicts that there are no triple couplings of the *Z* which drive the *Z* pair production. No *ZZγ* or *ZZZ* couplings are thought to exist. Hence the process is thought to occur by way of double *Z* radiation. Experimental data from LEP indicate that there are no anomalous couplings seen in this process to a level of sensitivity set by the number of observed events. The student might now use COMPHEP to predict the data shown in Fig. 1.6 and to examine the Feynman diagrams that COMPHEP provides for this process.

What about the predicted quartic couplings? The LEP facility at CERN has an energy that is insufficient to produce three heavy gauge bosons, so we have, as yet, no data

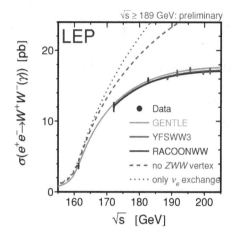

Figure 1.5   Data ([1], with permission) from the L3 experiment at LEP on the cross section for *WW* pair production in electron–positron annihilations. There is a *ZWW* coupling (Fig. 1.3) which is required to describe the data properly.

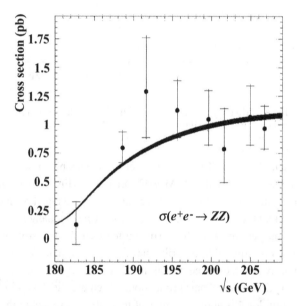

Figure 1.6   Data ([2], with permission) from the OPAL experiment at LEP on the cross section for *ZZ* pair production in electron–positron annihilations. The line indicates the SM prediction.

to check against the predicted quartic couplings except in the case where the third boson is a photon. The triple gauge boson final states are produced by way of diagrams, some of which contain quartic gauge boson couplings. The student should verify that assertion by looking at the Feynman diagrams for $e^+ + e^- \rightarrow W^+ + W^- + Z$ in COMPHEP.

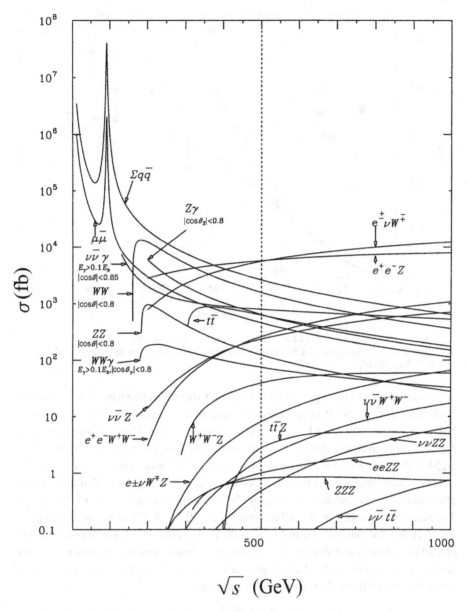

Figure 1.7 Cross sections ([3], with permission) in fb = 0.001 pb, for various processes as a function of CM energy in electron–positron annihilations. *WWZ* and *ZZZ* have quartic gauge boson contributions and cross sections at 1 TeV CM energy of ~100 fb and 1 fb respectively. The region with CM energy below 200 GeV has already been explored by the LEP experiments.

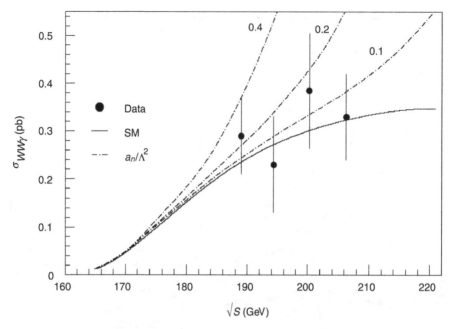

Figure 1.8   Cross section at LEP ([4] – with permission) for the production of the $WW\gamma$ final state as a function of CM energy. The dotted lines indicate models beyond the Standard Model.

The observation of these processes at the predicted cross section would be an important confirmation of the SM. However, the data taking awaits a decision to build a new energy frontier accelerator to extend the electron–positron collider CM energy range shown in Fig. 1.1. The proposed device is called the Linear Collider (LC). A CM energy of $>251 = 80 + 80 + 91$ GeV is needed to make the ZWW final state, as seen in Fig. 1.7.

Meanwhile, there are data from the final data-taking period at the LEP machine on the cross section for the production of the $W^+W^-\gamma$ final state as a function of CM energy. The expected cross section of ~ 0.3 pb compared to 20 pb for WW is indicated in Fig. 1.6. The fact that the data shown in Fig. 1.8 are in agreement with the Standard Model prediction indicates that this specific quartic gauge boson coupling appears to exist and have the predicted strength. That fact gives added support to the prediction that the weak gauge bosons are themselves carriers of weak charge.

1.6 The Higgs mechanism for bosons and fermions

We now turn to the Higgs boson as the last undiscovered SM particle. First we need to discuss further the weak interactions. They were parameterized by Fermi in the 1930s as an effective four fermion interaction with a universal coupling, $G \sim 10^{-5}$ GeV^{-2}. The parameter G is not dimensionless, so we expect that it is not a fundamental quantity. The muon decay width Γ_μ is, by dimensional argument (G defined so that the decay rate is proportional to G^2, $[G^2] = [1/M^4]$, $[\Gamma] = [M]$), proportional to the fifth power of the

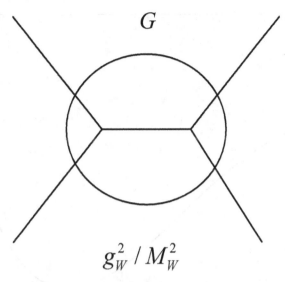

Figure 1.9 Schematic representation of the decomposition of the effective Fermi coupling constant G into a dimensionless coupling g_W and a propagator for a vector boson of mass M_W when the momentum transfer is much less than the W mass.

muon mass, $\Gamma_\mu \sim G^2 m_\mu^5$, which yields an estimate for the decay width Γ of $1/(6.6 \times 10^{-10}$ sec) or 0.66 nsec for the lifetime, τ. The decay width has units of mass, while the lifetime has units of time or inverse mass, $[\Gamma] = M$, $[\tau] = 1/M$. Since a strong process lifetime could be estimated to be $\tau \sim \hbar/\Gamma \sim (\hbar/\alpha_s m_\mu) \sim 10^{-22}$ s, the decays are indeed slow with respect to strong interaction rates.

The Fermi effective theory is not renormalizable. A first attempt at modification is to replace the four-fermion "contact" interaction with a "propagator" which spreads the interaction out in space-time and thus makes the interaction less singular. This is shown schematically in Fig. 1.9 for $q^2 \ll M_W^2$. We need to assign a large mass to the weak W boson in order to ensure that the interaction is weak at low energies. Effectively, then, $G \to g_W^2/M_W^2$. The fundamental strength of the weak interactions, g_W, then becomes comparable to the electromagnetic coupling e. Assuming $g_W \sim e = 0.303$, we then find that $1/\sqrt{G} = 296$ GeV or $M_W \sim g_W/\sqrt{G} = 89.7$ GeV.

This improves things but does not solve them. The apparent weakness of the weak interactions at low energies requires that the W and Z acquire masses ~ 100 GeV. However, we also need the theory to be a renormalizable one.

It turns out that simply adding a term to the fundamental Lagrangian with an explicit W mass term destroys the renormalizability of the theory. Therefore it is necessary, in the simplest case, to hypothesize the existence of a fundamental scalar field which has an interaction potential $V(\phi)$ shown in Eq. (1.8). The interactions represented by this potential induce the masses of the vector gauge bosons. The potential represents the self-coupling of the Higgs bosons and contains two arbitrary parameters. The parameter

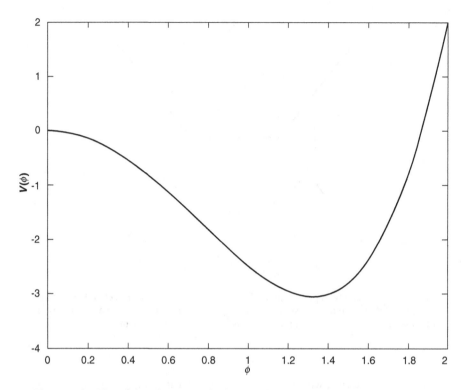

Figure 1.10   Plot of the Higgs potential for a particular choice of the two parameters
that define the Higgs interactions.

λ is dimensionless (see Appendix A), while the parameter μ has the dimension of mass.

$$V(\phi) = \mu^2 |\phi|^2 + \lambda |\phi|^4. \tag{1.8}$$

The minimum of the Lagrangian, $\partial V/\partial \phi = 0$, which we identify as the vacuum state, occurs not at zero field but at a non-zero "vacuum expectation value" $\langle \phi \rangle$:

$$\langle \phi \rangle^2 = -\mu^2/2\lambda. \tag{1.9}$$

In most other cases in physics the vacuum is a state with zero average field. However, a classical situation with similar phenomenology occurs in superconductivity, which may be familiar to the reader. The free massless photon acquires a mass inside a superconductor and thus the electromagnetic field is excluded from a superconductor (recall the exponential suppression of the potential for a massive boson) except for a small "skin depth" near the surface of the superconductor in the Landau–Ginzburg theory of superconductivity. We will see, by analogy, that it is the interaction of this vacuum Higgs field with all other fermions and bosons that endows them with a mass. A plot of Eq. (1.8) for a particular choice of μ and λ is shown in Fig. 1.10, which illustrates that the minimum of $V(\phi)$ occurs at a non-zero value of the field.

The alert reader will note that the Lagrangian density, $\ell \sim (\partial\overline{\phi})^*\partial\phi + V(\phi)$, does not vanish in the vacuum state. There is a "cosmological term," $V(\langle\phi\rangle) \sim \lambda\langle\phi\rangle^4$, which we will discuss in Chapter 6. This term implies that the vacuum state possesses an energy density, unfortunately far in excess of that observed in Nature.

Recall that the covariant derivative contains the fields W and Z. Suppose an additional field ϕ exists and has a vacuum expectation value. The quartic couplings we described already for the vector gauge bosons then give mass to the W and Z. This is called "spontaneous electroweak symmetry breaking" because the masses are not explicitly assigned initially but appear spontaneously by way of interaction with the Higgs vacuum field. The gauge replacement for the kinetic energy of the hypothesized scalar field leads to a weak boson mass $\sim g_W\langle\phi\rangle$, since the W mass term in the Lagrange density is $\sim M^2\overline{\varphi}_W\varphi_W$, where φ_W is the vector gauge field of the W boson.

$$(D\overline{\phi})^*(D\phi) \sim \left[g_W^2\langle\phi\rangle^2\right]\overline{\varphi}_W\varphi_W. \tag{1.10}$$

The weak gauge bosons, W^+, Z^0, and W^-, acquire a mass by interacting with the "vacuum expectation value" of the Higgs boson field, while the photon, γ, remains massless. The coupling g_W can be connected to G by noting that the four-fermion interaction can be related to the effective propagator, $G \sim g_W^2/M_W^2$, $g_W = e\sin\theta_W$. Thus, from G, e, and $\sin\theta_W$ we can predict M_W. The Weinberg angle in turn can be determined from neutral current weak neutrino interactions (see Appendix A). The resulting prediction, $M_W \sim 80$ GeV, was confirmed in the early 1980s at CERN in the proton–antiproton collider experiments, UA1 and UA2. The vacuum Higgs field thus has the experimentally determined value $\langle\phi\rangle \sim 174$ GeV.

$$M_W = g_W\langle\phi\rangle/\sqrt{2}, \ M_Z = M_W/\cos\theta_W. \tag{1.11}$$

The ratio of the W and Z masses is predicted, $M_Z = M_W/\cos\theta_W$ (see Appendix A). This prediction of the SM has also been experimentally established to high precision.

The W and Z masses are fixed by the Higgs mechanism and specify one of the two parameters of the Higgs potential. Let us turn now to fermions. The masses of the leptons and quarks range over five orders of magnitude from the electron, 0.5 MeV, to the top quark, 175 GeV (see Fig. 1.1). In the interest of economy, we again use the vacuum expectation value of the Higgs field to create the mass. A fermion mass can be induced using the Yukawa couplings of fermion pairs to the Higgs boson. These couplings are not specified by the gauge symmetry; they are simply put in by hand. This is convenient and compact, but does not lead to new predictions.

The Yukawa coupling, g_f, of the Higgs field to the fermions is postulated to be $\ell \sim g_f[\overline{\psi}\phi\psi]$. A vacuum expectation value for the Higgs field, $\ell \sim g_f\langle\phi\rangle[\overline{\psi}\psi] = m_f[\overline{\psi}\psi]$, then induces a fermion mass term, m_f. (See Appendix A.) The coupling of the Higgs to light quarks is rather weak with respect to coupling to W – in the ratio m_f/M_W.

$$m_f = g_f\langle\phi\rangle = g_f[\sqrt{2}M_W/g_W],$$
$$g_f = g_W(m_f/M_W)/\sqrt{2}. \tag{1.12}$$

We have not gained anything in predictive power, but the Higgs field can generate the masses of all the fermions just as it does for the gauge bosons. For each mass we have exchanged our ignorance of a mass for an unknown coupling constant, g_f. However, there is still the prediction that the Higgs boson couples to fermions with strength proportional to the mass of that fermion. Confirmation of that SM prediction is very important and will be looked for in future.

1.7 Higgs interactions and decays

In the previous section we saw how the vacuum expectation value of the Higgs field could give a mass to all the particles in the SM. The excitations, ϕ_H, of the Higgs field, $\phi \sim \langle \phi \rangle + \phi_H$, imply the existence of field quanta just as the excitations of the electromagnetic field are identified as the photon. The couplings of the Higgs excitation to the bosons and fermions are indicated schematically in Fig. 1.11.

There are interactions of the H particle both with gauge bosons and self-interactions, as was the case when we looked at the vector gauge couplings. Examining the kinetic energy term for the Higgs field, $\ell \sim (\partial \overline{\phi})^* \partial \phi$, and making the gauge replacement of the derivatives, $\partial - i g \phi$, there are triple and quartic couplings of the Higgs quanta to the electroweak gauge bosons. Therefore we expect $\varphi_W \overline{\varphi}_W \phi_H$, $\varphi_W \overline{\varphi}_W \phi_H \phi_H$ couplings in analogy to Eq. (1.7). The gluons and photons do not carry flavor. Hence they are "flavorblind," and do not couple directly to the Higgs.

We will defer any discussion of the Higgs self-interactions that are specified in Eq. (1.8). Suffice it to say that, as gauge couplings, they are specified by the gauge principle, just as those of the W and Z are. Therefore, they are a clear prediction of the SM and should be experimentally challenged.

The triple coupling is to the mass of the W and Z bosons, $\ell \sim g_W^2 \langle \phi \rangle [\overline{\varphi}_W \varphi_W \phi_H] \sim g_W M_W [\overline{\varphi}_W \varphi_W \phi_H]$. The existence of this interaction means that the Higgs scalar, if it is energetically possible, preferentially decays into W and Z pairs since those couplings are much stronger than the couplings to the fermions.

The decay width into W pairs is shown below. The rate depends on the weak fine structure constant and on β, where β is the $L = 0$ (L is the WW angular momentum) threshold factor $\sqrt{1 - (2 M_W/M_H)^2}$, which is the velocity of the W in the Higgs CM with respect to c.

$$\Gamma(H \to WW)/M_H \sim (\alpha_W/16)(M_H/M_W)^2 \beta. \tag{1.13}$$

The centrifugal suppression factor β^{2L+1} is due to the fact that larger angular momentum means larger centrifugal force, pushing the Ws away from the Higgs and reducing the decay probability. This factor is familiar from the study of the central force problem in quantum mechanics, for example the hydrogen atom.

Unfortunately, there are two parameters defining the Higgs potential, Eq. (1.8), and we have fixed only one by experimentally finding the vacuum expectation value of the field (see Appendix A, $G \sim \alpha_W/M_W^2 \sim 1/\langle \phi \rangle^2$). Thus the Higgs mass is an unknown parameter of the SM, which must be determined experimentally. Using the Higgs potential,

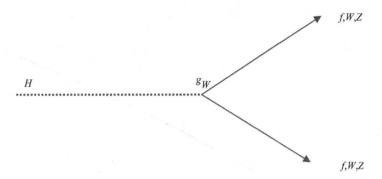

Figure 1.11 Schematic representation of the interactions of the Higgs boson with both fermions and bosons in the trilinear case.

$V(\phi)$, expanding about the minimum at $\phi = \langle\phi\rangle$, and identifying the mass term in ℓ as $M_H^2\bar{\phi}_H\phi_H$, we find that the mass is $M_H = \langle\phi\rangle\sqrt{2\lambda} = 246\,\text{GeV}\sqrt{\lambda}$. Since λ is an arbitrary dimensionless coupling, there is no prediction for the Higgs mass in the SM.

A rough upper limit for the mass can be inferred as the mass value when the Higgs excitation ceases to be a recognizable resonant state, which is when the weak interactions become strong.

$$\Gamma(H \to WW)/M_H \sim 1 \text{ if } M_H \sim M_W(4/\sqrt{\alpha_W}) \sim 1.7\,\text{TeV}. \qquad (1.14)$$

We move now to the coupling of the Higgs to fermions, which is defined by the Yukawa coupling with a fermion coupling constant, g_f. Therefore the Higgs couples to fermions proportional to their mass (Eq. (1.12)). The very low mass, ~ 4 MeV, of the u and d quarks which make up the proton, which is the particle we will collide with itself or its antiparticle, means that the Higgs boson couples very weakly to ordinary matter. The coupling is $g_u \sim 0.000\,023$, very weak compared to $e = 0.303$, $g_W = 0.65$ and $g_s = 1.12$. Gluons are not directly coupled either. This weak coupling makes discovering and measuring the properties of the Higgs scalar a great experimental challenge. In contrast, the heaviest quark, the top, is strongly coupled, $g_t \sim g_W\,(m_t/M_W)/\sqrt{2} \sim 0.99$. The Higgs decay width into quarks is shown in Eq. (1.15). For leptons the same result holds save that the color factor of three should be omitted as we no longer sum over all final state colors. The decay is into a fermion–antifermion pair which has the quantum numbers $P =$ parity, $L =$ orbital angular momentum, $S =$ spin angular momentum, and $J =$ total angular momentum. The pair has charge conjugation C and parity P; $C = (-1)^{L+S}$, $P = (-1)^{L+1}$. The Higgs is a scalar, $J^{PC} = 0^{++}$, so that the pair must have $L = 1$, because the intrinsic parities of a quark and an antiquark are opposite. The threshold factor mentioned above is, for $L = 1$, β^3.

$$\Gamma(H \to q\bar{q})/M_H \sim (3\alpha_W/8)(m_f/M_W)^2\beta^3. \qquad (1.15)$$

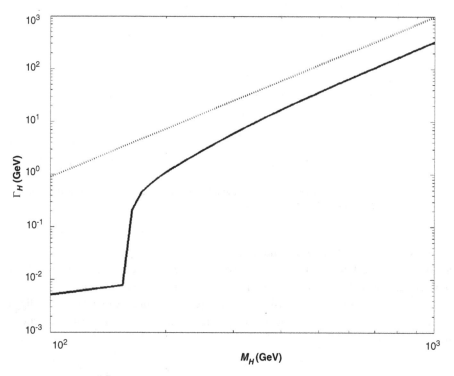

Figure 1.12   Higgs decay width as a function of mass summed over all fermion and boson final states. The dotted line indicates a cubic dependence of the width on the Higgs mass.

The total Higgs decay width as a function of Higgs mass is given in Fig. 1.12. Note the M^3 behavior at high masses, as expected due to the dominance of the WW and ZZ decay modes. At low masses, a linear dependence of the decay width on Higgs mass into quarks is expected, from Eq. (1.15), and is seen as a steep drop in width with decreasing Higgs mass. The experimental mass resolution expected in LHC experiments (Chapter 5) is much larger than the intrinsic width of the Higgs at low mass. Thus, the total width is dominated by the experimental mass resolution and the intrinsic width will be unobservable. Clearly, if the Higgs is a relatively low mass object, optimizing the detector mass resolution will be of critical importance.

The ZZ and WW widths can be computed in COMPHEP and compared to Fig. 1.12. The student is encouraged to see if the results can be duplicated. The COMPHEP program also allows us to evaluate the "off shell" decays of a Higgs into $ZZ^* = Z\ell^+\ell^-$ which can occur at a mass below $2M_Z$ because of the spread in mass of the unstable Z resonance characterized by the Breit–Wigner width (see Appendix A).

Note the threshold behavior at Higgs mass equal to about twice the W mass and the ultimate, high mass cubic dependence on the Higgs mass. Note also that a 1 TeV mass Higgs has a ~ 0.3 TeV decay width into $ZZ + WW$ pairs, so that the width to mass ratio is already 30%. The Higgs branching ratio into top pairs is smaller than that into W or Z pairs, and is ignored in this estimate.

We will return to the subject of finding the Higgs in Chapter 5 after we arm ourselves with the tools we need in the next three chapters.

1.8 Questions unanswered by the SM

We have tried in this first chapter to give an overview of the accumulated wisdom in high energy physics obtained over the last 40 years or more. The treatment has been brief and the mathematics has been simplified. Nevertheless, we hope that the basic insights of the Standard Model have been presented and partially explained so that the student has a theoretical context for the SM in place. We also assume that the student has by now acquired some facility with the COMPHEP program and will reproduce the examples given in the text as the exposition unfolds.

There are many arbitrary parameters contained in the Standard Model, for example, the three fine structure constants, α, α_s, α_W, the six masses of the quarks, and the three masses of the leptons (six if neutrinos are allowed to have small masses). Many of these parameters have to do with the replication of the pattern in the Standard Model into three generations. We do not yet understand why they take the values we measure experimentally.

There are many other experimental facts that are simply put into the SM "by hand" because the fundamental reason for them is not yet understood. For example, charge quantization is imposed; all electric charges, Q, appear in $1/3$ units of the electron charge e. Proton stability is put in by hand; there is no fundamental dynamical reason known why protons do not decay. In contrast, "color" and charge are associated with an exact symmetry for the strong and electromagnetic interactions. Thus we expect charge and "color" to be conserved rigorously.

There are observed to be three "generations" of quarks and leptons, as indicated schematically in Fig. 1.2. The reason for the existence of three and only three "generations," distinguished only by a "flavor" quantum number such as strangeness (s), charm (c), beauty (b), or top (t) is unknown.

The charge changing (beta decay) weak interactions, mediated by the charged W bosons, do not conserve flavor. Thus, the heavy quarks and leptons ultimately decay to the u, d and e familiar to us as the constituents of ordinary matter. The most likely charge changing quark transitions are contained within a generation; $u \to d + W^+$, $c \to s + W^+$, $t \to b + W^+$. The strength of these charge changing quark transitions is nearly the same as the strength of the charge changing lepton transitions, $e^- \to \nu_e + W^-$, $\mu^- \to \nu_\mu + W^-$, $\tau^- \to \nu_\tau + W^-$, embodied in the universal Fermi decay constant G. The favored quark and lepton transitions can be viewed as a downward transition in Fig. 1.2 with accompanying W emission.

As discovered in the 1970s, there are also neutral weak interactions mediated by the Z^0. There are no flavor changing neutral weak interactions by construction; they are required to be "diagonal" in flavor. For example, there are no $c \to u + Z^0$ transitions allowed. The Z boson decays into flavor pairs of quarks and leptons but, for example, $Z^0 \to c + \bar{u}$ is not allowed, nor are $\mu^+ + e^-$ decays. In Fig. 1.2 there are no "horizontal" neutral weak transitions. Another example of a forbidden transition in the SM is $\mu \to e\gamma$, which is not allowed because flavor is not conserved and charge does not change. The experimental upper limit of the muon decay probability into this final state is 2×10^{-11}, which is indeed small.

We list below some of the unresolved fundamental questions that are not answered in the context of the SM. It would be the height of presumption to imagine that we can do more than explain the experimental program, which is now being mounted to explore the second question, to which we devote Chapters 2–5 of this text. We will, however, very briefly return to these questions in Chapter 6. Our aim here is to bring these questions forward to the student so that he or she is aware that the SM, although a wonderful edifice which explains all our present experimental data, appears to be incomplete and therefore unsatisfying. Clearly, there remains a lot of work for the next generation of high energy physicists to do!

Questions

1. How do the Z and W acquire mass while the photon does not? (Chapter 1)
2. What is M_H and how do we measure it? (Chapters 4,5)
3. Why are there three and only three light "generations"? (Chapter 6)
4. What explains the pattern of quark and lepton masses and mixing?
5. Why are the known mass scales so different?

    $\Lambda_{QCD} \sim 0.2$ GeV (strong interaction field)

    $\ll \langle \phi \rangle \sim 174$ GeV (electroweak scale)

    $\ll M_{GUT} \sim 10^{16}$ GeV (Grand Unified scale)

    $\ll M_{PL} \sim 10^{19}$ GeV (Planck mass scale where gravity becomes strong).

6. Why is charge quantized?
7. Why do neutrinos have such small masses?
8. Why is matter (protons) approximately stable?
9. Why is the Universe made wholly of matter? (CP violation)
10. What is "dark matter" made of? There is no plausible SM candidate particle. What is "dark energy"?
11. Why is the cosmological constant so small? The vacuum Higgs field leads to a constant which is 10^{55} times the closure density of the Universe.
12. How does gravity fit in with the strong, electromagnetic, and weak forces?

Exercises

1. Download the COMPHEP code and read the Users' Manual.
2. Read the worked example in Appendix B. Find the cross section for electron–positron production of W pairs at 200 GeV and compare it with the result quoted in the text, Fig. 1.4.
3. Download the .pdf reader from the Adobe site quoted in the introduction.
4. Use your web browser to find the Fermilab publications site, fnalpubs.fnal.gov. Then click on preprints and search. Look for author "Montgomery" and find "The Physics of Jets". Download the paper as a .pdf file. Then go to the

site fnalpubs.fnal.gov/archive/1998/conf/Conf-980398.pdf. Compare with [8] of Chapter 2, H. Montgomery, Fermilab – Conf-98–398 (1998).

5. Evaluate the Fourier transform of the Yukawa potential and verify that it has the form of a "propagator" with mass as indicated in Eq. (1.6).

6. Use COMPHEP to find the cross section for electron–positron production of ZWW and compare the result, at 1 TeV CM energy, with that shown in Fig. 1.6.

7. Find the minimum of the Higgs potential, Eq. (1.10), to confirm Eq. (1.11).

8. Evaluate the Higgs width into W pairs for a 1 TeV Higgs boson.

9. Evaluate the Higgs width into b quark pairs for a 120 GeV Higgs boson.

10. Use COMPHEP to evaluate the widths given in Exercises 8 and 9 and compare the results.

11. If the proton had a lifetime of 10^{31} years, how many decays would occur in your body in a 1-year period?

12. If the neutrino to proton ratio in the Universe is $\sim 10^9$ and if the mass density of the Universe is $\sim 1 \; p/m^3$, estimate the neutrino mass needed if they are to be responsible for the entire mass density.

13. Use COMPHEP to look at electron–positron production of $H + Z$. Check the Feynman diagrams. For a Higgs mass of 130 GeV find the cross section at a CM energy of 250 GeV. What is the cross section for $H + H + Z$ at an energy of 500 GeV? Look at the Feynman diagram to confirm that triple H and quartic H couplings contribute to this latter process.

14. Look at the COMPHEP model parameters for quark and lepton masses and compare with Fig. 1.1.

15. Use COMPHEP in the SM and compare the list of particles with that given in Fig. 1.2.

16. Find the W and Z decay width and branching fractions in COMPHEP, $W \to 2*x$, $Z \to 2*x$. Compare with the data shown in Chapter 4.

17. Use COMPHEP to look at electron–positron W pair production. How many Feynman diagrams are there? Turn all but one off and evaluate each in turn. Which is largest? What is the full cross section? Are there destructive interferences? Look at the energy dependence of each diagram too. In particular, show that with only the neutrino exchange diagram active the cross section at a CM energy of 200 GeV is ~ 43 pb.

18. Use COMPHEP to find the cross section at 1 TeV CM energy for electron–positron production of WWZ. Check the Feynman diagrams to see that this process probes quartic gauge boson self-couplings.

19. Use COMPHEP to explore the vertices in the Lagrangian of the SM and compare with the results quoted in this Chapter and "derived" in Appendix A.

References

1. Bethke, S., MPI-PhE/2000-02.
2. OPAL Collaboration, CERN-EP/2003–049, July (2003).
3. *Linear Collider Physics*, Fermilab – Pub-01/058-E, May (2001).
4. L3 Collaboration, CERN – EP/2001–080.

Further reading

Aitchison, J. R. and A. J. G. Hey, *Gauge Theories in Particle Physics*, 2nd edn. Philadelphia, Adam Hilger (1989).

Bjorken, J. D. and S. D. Drell, *Relativistic Quantum Fields*, New York, McGraw-Hill (1965).

Cottingham, W. and D. Greenwood, *An Introduction to the Standard Model of Particle Physics*, Cambridge, Cambridge University Press (1998).

Gottfried, K. and V. Weisskopf, *Concepts of Particle Physics*, Vol. II, New York, Oxford University Press (1986).

Green, D., *Lectures in Particle Physics*, Singapore, World Scientific (1994).

Halzen, F. and A. D. Martin, *Quarks and Leptons*, New York, John Wiley (1984).

Particle Data Group, Review of particle properties, *Phys. Rev. D. Particles and Fields* **50**, August 1 (1994).

Quigg, C., *Gauge Theories of the Strong, Weak, and Electromagnetic Interactions*, Reading, Massachusetts, Benjamin/Cummings (1983).

2

Detector basics

Facts are stubborn things; and whatever may be our wishes, our inclinations, or the dictates of our passions, they cannot alter the state of facts and evidence.

John Adams

When you can measure what you are speaking about, and express it in numbers, you know something about it.

William Thomson

2.1 SM particles – mapping into detector subsystems

Chapter 1 served to define the particle content and interactions of the Standard Model (SM). The discussion of the Higgs boson width in Chapter 1 also showed that detector resolution would determine the sensitivity of searches for low mass Higgs particles. Our plan is to discuss in this chapter how the fundamental particles of the SM are detected and their kinematic properties measured. Specifically, we want to discuss the accuracy that we can expect to achieve in measuring the vector position and momentum of each SM particle that is produced in a collision.

We also wish to do "particle identification," that is, to identify a produced particle unambiguously as a unique element of the "periodic table" of the SM, which was shown in Fig. 1.2. We will use that information in the later chapters because it will inform on the optimal search strategies for new particles.

The discussion of detection principles that is given here will be very schematic. Several references are given at the end of this chapter, which supply many details of potential interest to the student. We assume that the reader is familiar with magnetic fields, ionization energy deposit in materials, and the electromagnetic interactions of charged particles.

A schematic view of a typical general purpose detector used in high energy physics experiments is shown in Fig. 2.1. The detector itself is logically broken into distinct subsystems. A solenoid electromagnet coil produces a large volume of axial magnetic field, in this example of strength 4 T (1 T = 1 Tesla = 10 000 Gauss). The purpose of this magnetic field is to bend all of the charged particles, which are emitted from the production point, or production vertex, by an amount that depends on the momentum and sign of the charge of the produced particles.

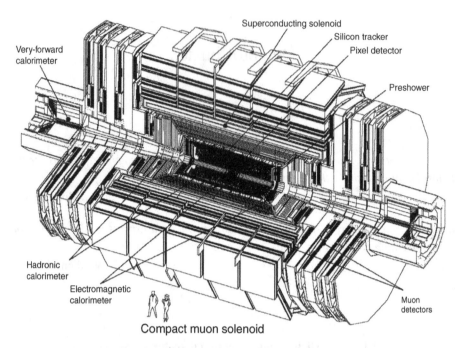

Figure 2.1   A general purpose detector used in proton–(anti)proton collider experiments. The subsystems used are: a tracking system, a hermetic calorimeter system which is subdivided into an electromagnetic (ECAL) and a hadronic (HCAL) section, a large solenoid magnet coil to provide a large volume filled with magnetic field, and the iron needed to supply the magnetic flux return for the magnet. The flux return is itself instrumented with chambers to measure the trajectories of the muons ([1] – CMS, with permission).

A measurement of the trajectories of the charged particles then results in the determination of their position and momentum vectors. The ionization energy loss in the tracking detector elements is small. Therefore, this detection device is not "destructive" of the properties of the particle. In turn, that means we can make subsequent redundant measurements of, say, the particle energy as it escapes from the production vertex.

Working our way out from the interaction point at increasing distances we exit the tracker and next encounter electromagnetic calorimetry followed by hadronic calorimetry. The purpose of the calorimetric detectors is to measure the energy of both the charged and neutral particles, which are incident upon it. These detector systems extend down to angles of about $0.8°$ to the incident beam directions. They are the two main longitudinally, or depth segmented, "compartments" of the calorimetry.

The electromagnetic calorimeter initiates the interaction of photons and electrons. Recall that these fundamental particles have only electromagnetic and weak interactions. The hadronic calorimeter elements initiate the interactions of all the strongly interacting particles, such as quarks and gluons, or, more accurately, their "decay" products. By totally absorbing the energy of the incident particles and by sampling that absorbed energy, the calorimetry makes a measurement of the energy of almost all the produced particles.

Finally, the muons, which have only electromagnetic and weak interactions, are detected and identified in tracking chambers embedded in the magnetic return yoke of the magnet. The muons have the same interactions as electrons ("who ordered that?"), but they are about 200 times heavier. Therefore, they do not radiate significantly at the energies considered here and only lose energy by ionization. When all other particles have been absorbed what remain are the muons.

Comparing the initial energy transverse to the proton and (anti)proton beams (E_T is approximately zero) and the detected transverse energy of all particles in the final state, we can look for a mismatch. Any missing energy implies either a mis-measurement, incomplete detector coverage, or that neutrinos, which interact only weakly, were produced and escaped detection. We consider only transverse energy imbalance because energy can escape undetected near to the vacuum pipe containing the beams, which means that the final state total longitudinal energy is poorly measured.

The accuracy of the measurement of the momentum, P, or energy, E, of single particles is defined by the resolution of the tracking detectors in the magnetic field or the calorimetric energy resolution. In both cases the resolution is represented by expressions containing two terms for the fractional error, which are "folded in quadrature" (that means $a \oplus b = \sqrt{a^2 + b^2}$). The resolution for tracking, dP/P, has a term that increases with momentum, while the resolution for calorimetry, dE/E, has a term that decreases with energy. If the b and d factors can be ignored, this different behavior of the energy resolution makes calorimetry the detector of choice at very high energies.

$$dP/P = cP \oplus d,$$
$$dE/E = a/\sqrt{E} \oplus b. \tag{2.1}$$

The tracking resolution has a term due to the finite accuracy of the measurements of the deflection angle of the particle in the magnetic field, c, and a term due to multiple scattering, d. The calorimetric terms are due to stochastic fluctuations in the sampled energy, a, and non-uniformity of the medium, b. Examples will be given later in this chapter in order to set the numerical scale.

In Chapter 1 we provided a table (Fig. 1.2), which defined all the fundamental particles of the Standard Model except the Higgs boson. For purposes of detection, we will now separate them into strongly interacting particles, electromagnetically interacting particles, and weakly interacting particles.

The strongly interacting particles are gluons (g) and quarks (u, c, t, d, s, b). The particles with electromagnetic interactions are photons and charged leptons (γ, e, μ, τ). The weakly interacting particles are the EW gauge bosons, W and Z, and the neutrinos, ν_e, ν_μ, ν_τ. Strictly speaking the neutrinos are not directly detected. Their presence in the final state is inferred from the existence of "missing" transverse energy, which means that the sum of all transverse energy in the final state is substantially different from zero.

Table 2.1 *Fundamental elementary particles in the Standard Model, their detection in particular detector subsystems, and a signature allowing for particle identification in those subsystems.*

Particle	Signature	Detector
$u, c, t \rightarrow W + b$ d, s, b g	Jet of hadrons (λ_0)	Calorimeter
e, γ	Electromagnetic shower (X_0)	Calorimeter (ECAL)
ν_e, ν_μ, ν_τ $W \rightarrow \mu + \nu_\mu$	"Missing" transverse energy	Calorimeter
$\mu, \tau \rightarrow \mu + \nu_\tau + \bar{\nu}_\mu$ $Z \rightarrow \mu + \mu$	Only ionization interactions dE/dx	Muon absorber
c, b, τ	Decay with $c\tau > 100$ μm	Silicon tracking

This separation, which is made according to the strongest force felt by the SM particle, is the first part of particle identification.

Basically, the calorimetry does a large part of the energy measurement of all the particles, as seen in Table 2.1. The electromagnetic compartment of the calorimetry gives us electron and photon energies and positions (specified by independently recorded polar and azimuthal angular "pixels"), while the hadronic compartment gives us the position and energy of the quarks and gluons. The characteristic that allows us to separate hadrons and electrons is achieved because of the large difference in mean free path for electromagnetic interaction, the radiation length X_0, and that for hadronic interaction, λ_0. For lead, the ratio is about 1:30.

The muons are uniquely identified as those charged particles that have only ionization interactions and thus penetrate deeply into the steel return yoke. The detectors in the yoke serve the purpose of doing muon particle identification.

The last row in Table 2.1 requires further explanation. Silicon detectors can now easily be constructed with a separation between detection elements, or "pitch," of about 50 μm. Therefore, particles that are produced at the primary interaction vertex and subsequently weakly decay at a secondary vertex point can be detected and identified if the distance between the primary and the secondary vertices exceeds about 10–100 μm. SM particles of this type include the c quark, the b quark, and the τ lepton.

Let us estimate the decay width of a c quark to an s quark in the specific decay, $c \rightarrow s + e^+ + \nu_e$. This is a decay within a generation, so we expect that the mixing matrix element is ~ 1. The decay can be visualized as first the emission of a virtual W, $Q \rightarrow q + W$, which then virtually decays into $\ell + \nu$. The two distinct vertices mean that the Feynman amplitude is proportional to the weak fine structure constant, while the decay width is proportional to the square. The virtual W propagator leads to $1/M_W^4$ behavior. Thus, by dimensional argument we expect scaling as the fifth power of the parent mass

Table 2.2 *Particle identification in a general purpose detector.*

Particle type	Tracking	ECAL	HCAL	Muon
γ				
e				
μ				
Jet				
E_t **miss**				

(see Chapter 4).

$$\Gamma \sim \alpha_W^2 (m/M_W)^4 m,$$
$$\Gamma \sim 2 \times 10^{-10}\,\text{GeV}. \tag{2.2}$$

Taking the mass of the c quark to be equal to 1.5 GeV (Fig. 1.1), we can very roughly estimate the charmed quark lifetime τ, and decay width Γ. The proper decay distance, $c\tau$, is estimated to be ~ 1.0 μm.

$$\tau = h/\Gamma,$$
$$c\tau \sim 1\,\text{μm}. \tag{2.3}$$

Therefore, we now understand why only the charm quark, the b quark, and the τ lepton appear in the last row of Table 2.1. The heavy quarks and leptons can be identified by finding resolvable decay vertices made available in a tracking volume extending over distances ~ 1 m. The decays shown in Table 2.1 for the top quark, the W, and the Z happen very rapidly with unresolved production and decay vertices.

The lighter unstable quarks and leptons (e.g. s quarks, muons) can be considered to be quasi-stable in that they have typical decay distances that are larger then the detectors themselves. For example, the muon is unstable but has a 2.2 μs (660 m) lifetime, so that it is very unlikely to decay before it exits the "generic" detector shown in Fig. 2.1. Therefore, we have SM particles that decay almost immediately, that decay within the tracker, and that decay outside the detector.

Particle identification at a more incisive level can often be accomplished by combining the information available from different subsystems of a general purpose detector. The principle is illustrated in Table 2.2. For example, electrons and photons both give energy deposits localized in the electromagnetic calorimeter. However, the charged electron has an associated track in the tracking subsystems, while the neutral photon does not ionize and leaves no track. Combining tracking and calorimetry therefore allows us to distinguish between electrons and photons. Muons, quark and gluon jets, and neutrinos all have unique signatures in a general purpose detector as seen in Table 2.2. Heavy quarks and leptons, b, c and τ, have, in addition, distinguishable secondary decay vertices.

Combining the information from the different detector subsystems is not only useful in particle identification but also in forming "triggers." Triggering, or pre-selecting events of interest prior to storing them on some permanent medium such as magnetic tape, is of primary importance in data taking at proton–(anti) proton colliders. The volume of data generated by a contemporary detector is enormous. There are millions of independent electronic channels recording data about an interaction and there are a billion interactions per second. Clearly, only a minuscule fraction of this information can be stored permanently. The rest must be discarded for all time. Given that perhaps only one hundred interactions per second can be stored for later study, we must quickly pick out one interaction in every ten million. Therefore we must be extremely careful and very sure that we choose the desired needle in the enormous haystack. Even so, the remaining volume of saved data is very large.

2.2 Tracking and "b tags"

We now go back and look in a bit more detail at the main detector subsystems. The tracking detectors may consist conceptually of a series of concentric cylinders for a typical collider detector. This geometry is often chosen with solenoid magnet coils that create axial magnetic fields, because then the particle trajectories are circles in the azimuthal or (r, ϕ) plane. At the very high interaction rates that will be required to search for the Higgs particle, detectors with the best possible rate capability will be needed. An example of such a detector, consisting of silicon pixels followed by silicon strips, is shown in Fig. 2.2. As we can see from the figure, the detectors are in fact built up by approximating a cylinder using small planar detectors oriented appropriately.

A major issue for the tracking detector subsystem is the efficient detection of the ionization energy left by charged particles, with a good signal to noise ratio so that spurious signals due to noise pulses are rejected. Spatial accuracy is obviously of the highest importance. Also important is the relative alignment of all of the planar elements making up the complete detector. A sufficient number of measurements of the position of the trajectory of the particle at different radii is needed to "pattern recognize" the helical path taken by a particle in the magnetic field and then "reconstruct" the track in space. The result of the tracker measurements is ideally a fully efficient determination of the vector position and momentum of all the charged particles emitted in the interaction but with no spurious tracks "found."

For each track we are measuring the bend angle, α, the angle by which the momentum vector is rotated, or "bent," in the magnetic field. The sense of the rotation tells us the sign of the charge of the particle. This angle is inversely proportional to the particle momentum, $\alpha \sim 1/P$. Thus, the fractional momentum error has a term due to angular

Figure 2.2 A photo of the mechanical prototype of a tracking system constructed entirely of planar silicon detectors. Concentric cylinders of detecting elements are built up out of identical rectangular sub-assemblies (CMS photo – with permission).

error $d\alpha$ which is proportional to the momentum (see Eq. (2.1)).

$$d\alpha \sim dP/P^2,$$
$$dP/P \sim (d\alpha)P = cP. \tag{2.4}$$

The additional term, which is folded in quadrature in Eq. (2.1), is due to multiple scattering, which is only important at low momentum. Since we are mostly interested in high transverse momentum physics, this term will be ignored from now on.

The bend angle increases with increasing magnetic field, $\alpha \sim B$, and the error on the bend angle decreases with improved spatial resolution. Therefore, there are basically two distinct strategies that can be employed to improve the momentum measurement made by a tracking detector: increase the field or improve the spatial resolution. At the present time, a 4 T field and a spatial resolution of a few μm, as afforded by silicon detectors, is at the technological limit. These precision tracking detectors operated in high fields have good momentum resolution. Typically, a 100 GeV particle will have its momentum measured at the one percent level.

Another important task performed by a tracking subsystem is the identification and measurement of secondary vertices. As we saw in Chapter 1, the Higgs is postulated to couple to mass. Therefore, detection of heavy quark and lepton decays is an important ingredient in Higgs searches. These heavy objects are unstable and decay weakly into lighter quarks and leptons respectively.

The lifetimes in the particle rest frame, in distance units, of the charm quark, the *b* quark, and the τ lepton are:

$$
\begin{array}{ll}
c\tau \sim (124 - 320)\ \mu\text{m} & c\ \text{quarks} \\
 \sim (468 - 495)\ \mu\text{m} & b\ \text{quarks} \\
 \sim 87\ \mu\text{m} & \tau\ \text{leptons.}
\end{array}
\tag{2.5}
$$

The quoted range of lifetimes for *c* and *b* quarks has to do with the fact that the decays of quark–antiquark bound states with large binding energy corrections due to the strong force are, in fact, what are measured and not the "bare" heavy quark decays. An isolated colored quark cannot exist, by assumption, so that it is the colorless bound states of quarks that are measured. The lifetime spread decreases for the *b* quark since it is about three times heavier than the *c* quark and higher mass means weaker strong interaction corrections (see Appendix A and Chapter 6) since the strong interactions become weak at high mass scales.

We saw in Chapter 1 that the weak interaction was responsible for the decay of the second and third generation quarks and leptons. The decay width for typical decay modes as a function of the available center of mass (CM) energy is given in Fig. 2.3, as is the spin correlation induced by the V-A nature of the weak interaction (see Chapters 4 and 5). The thick arrow in Fig. 2.3(b) indicates spin direction here while the thin arrow shows the momentum direction. We simply assert that the weak interaction imposes the constraint that particles have negative helicity, or spin anti-parallel to momentum, while antiparticles have positive helicity. The "generic" decay is of a heavy quark Q to a light quark q, lepton and antineutrino, $Q \rightarrow q + \ell^- + \bar{\nu}_\ell$.

Entries to Fig. 2.3(a) include the transitions between up and down quarks in free neutron beta decay and in charged pion decay. Other entries are the transitions between strange and up quarks, charm and strange quarks, and bottom and charm quarks. Where relevant, the legend in Fig. 2.3 shows the approximate square of the mixing matrix element for the particular quark decay, $V_{qq'}^2$, in terms of powers of the Cabibbo angle θ_c (see Chapter 6). The line given in the figure represents the fact that the decay width is closely proportional to the fifth power of the available energy over about fifteen orders of magnitude in the decay width.

In the rest frame of an unstable particle, where the proper time is labeled as t', there is a characteristic lifetime τ, as seen in Eq. (2.6). The time observed by the laboratory clocks is t, and $N(t)$ is the number of particles that survive at time t:

$$
\begin{aligned}
N(t') &= N(0)e^{-t'/\tau} \\
t &= \gamma t' = R/v \\
N(t) &\sim e^{-t/\gamma\tau} \sim e^{-Rm/Pc\tau}.
\end{aligned}
\tag{2.6}
$$

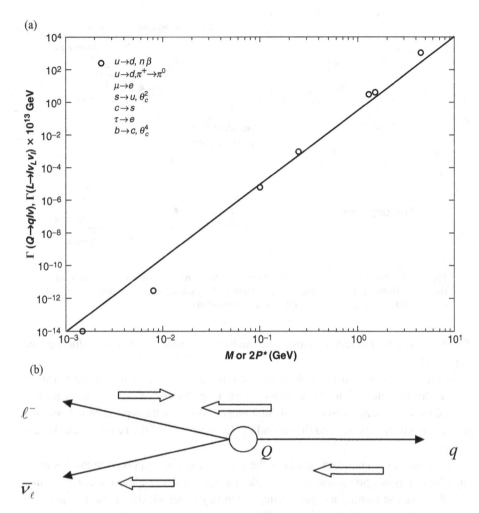

Figure 2.3 (a) The weak interaction decay width as a function of the available center of mass energy. The up and down quarks, the strange quark, the charm quark, and the *b* quark follow a single curve (the fifth power of *m*), as do the muon and the tau leptons. The strange and *b* quark decay widths are adjusted by the square of the quark mixing matrix elements (see Chapter 6) (b) Helicity structure of $Q(-1/3) \rightarrow q(2/3) + l^- + \bar{\nu}_l$ decays induced by the V-A weak interactions that make particles left handed (negative helicity) and anti-particles right handed (positive helicity). The direction of the momentum is indicated by the thin arrow, the spin direction by the thick arrow.

We use the relationships found in special relativity that the energy E and the rest mass m are related by $E = \gamma m$, where $\gamma = 1/\sqrt{1-\beta^2}$. The momentum P and energy E are related to the velocity, v, with respect to c, $\beta = v/c = P/E$ (see Appendix C). The total distance traveled before decay is R, so that $R = vt$. In the detector frame, the measured time t is dilated. The heavy quarks and leptons have mean decay distances of $\langle ct \rangle = c\tau\gamma$.

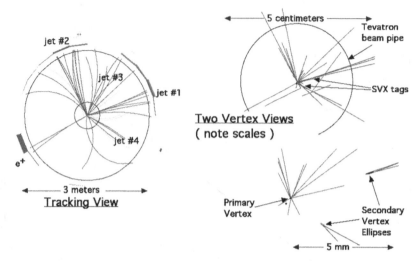

Figure 2.4   An axial view of a multi-jet event in the CDF detector. At a scale of 1 mm, the vertex from which the particles are emanating is resolved into a primary vertex and two secondary decay vertices (CDF – with permission).

Since $\gamma > 1$, silicon detectors with a strip pitch of ~50 μm or smaller are sufficient (see Eq. (2.5)).

In Fig. 2.4 we see an example from the CDF detector operating at the Fermilab accelerator complex. Note the ability of a tracking detector using silicon to resolve secondary vertices. At a distance scale of 5 mm or 5000 μm, the separation between the primary production vertex and the secondary decay vertices of the heavy quarks is very evident.

The identification of heavy quarks in the final state is very important in many studies of collider physics processes. For example, top quarks decay almost exclusively into $b + W$. If we can identify a b quark using secondary vertex identification (this is called "b tagging"), then we have taken a big step toward identifying the top quark.

2.3 EM Calorimetry – e and γ

The next detection subsystem that a particle encounters in exiting from the production point is the electromagnetic calorimeter. The two basic characteristic radiative processes, which create an electromagnetic "shower," are Bremsstrahlung radiation by the electrons and electron–positron pair production by the photons. There is a characteristic length scale for radiative processes in the material of the calorimeter called the radiation length, X_0. For example, X_0 is 0.56 cm in lead. Since an electromagnetic shower is initiated and runs its course in about twenty radiation lengths, or 11.2 cm in lead, an electromagnetic calorimeter can be quite compact.

There is a characteristic energy, which defines the termination of the electromagnetic shower multiplication processes. This is the critical energy, which is the energy

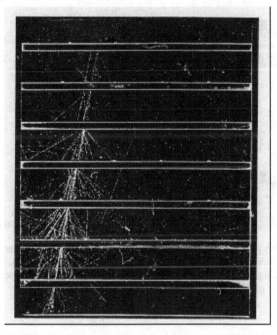

Figure 2.5 A photograph of the development of an electromagnetic shower in Pb plates. The number of particles in the shower builds up geometrically. After reaching a maximum, the shower then slowly dies off due to ionization loss ([2] – with permission).

below which radiative processes largely cease and particles in the shower lose energy only by ionization or other non-radiative processes. At this depth in the shower, called "shower maximum," the number of particles in the shower is a maximum and all have approximately the same energy, the critical energy. Given the absence of further particle production, the particles in the shower then lose energy and eventually come to rest.

For typical materials used in electromagnetic calorimeters, the critical energy, E_c, is approximately 2.5 MeV. Assuming that all particles in the shower share the energy equally, a 1 GeV electron incident on the calorimeter becomes, at the shower maximum, a shower of 400 particles, $N \sim E/E_c$. The stochastic fluctuation on the number of particles in the shower, N, then leads to an estimate for the fractional energy error of $\sim 5\%$, $(E = E_c N, \mathrm{d}E = E_c \sqrt{N}, \mathrm{d}E/E = 1/\sqrt{N})$.

A picture of a shower developing in sequential lead plates is shown in Fig. 2.5. The shower begins in the first two plates, reaches a maximum and then begins to die off.

There is a characteristic transverse size of a shower, also roughly X_0. This means that photons and electrons can be well-localized transverse to the point of impact on the calorimeter by the calorimetric measurement. Thus, the calorimetric technique measures both energy and position, although the position measurement is crude compared to tracking data.

There are several types of calorimetric signal readout. In Fig. 2.5 we saw the "sampling" type of calorimeter where the shower develops in passive heavy element plates

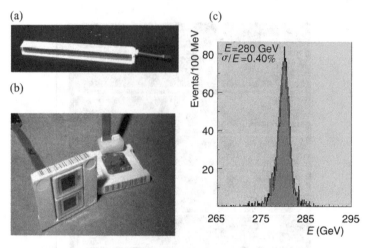

Figure 2.6 (a) A photograph of a fully active crystal electromagnetic detector. The emitted light from these crystals is detected in semiconductor elements, (b), and converted to an electrical signal, which is then recorded (c). This device is extremely accurate in its measurement of energy, (c). (CMS photo – with permission).

and is then sampled in gaseous or other low atomic weight active detector layers. Another type of readout in this fully active case is shown in Fig. 2.6. Typically, transparent scintillating crystals are used, which incorporate heavy elements. The light that is produced is then read out by a photon transducer of some sort. In principle, this is the most precise method of calorimetric energy measurement because there are no inactive materials with their attendant fluctuating unsampled energy deposits.

As seen in Fig. 2.6, at an energy of 280 GeV a fractional energy measurement of 0.4% is possible. Thus, electromagnetic calorimetry can have high precision, comparable even to that afforded by the tracking at energies above about 100 GeV.

In Equation (2.1) we defined the two parameters going into a calorimetric energy measurement. There was a "stochastic term," which is due to statistical fluctuations in the sampled energy of the shower and a "constant term," due to inhomogeneities in detector construction, which both contribute to the fractional energy error. For electromagnetic calorimetry, a stochastic coefficient of 2 percent, if the energy is expressed in GeV, and a constant term of 0.25 percent are at the present technological limit.

As we will discuss below, calorimeters are normally segmented into "pixels" which are limited in both polar and azimuthal angle. Each "pixel" functions independently and is read out as a distinct piece of information characterizing the interaction of interest. The variable used for equal spatial segmentation is not the polar angle but a quantity called the pseudorapidity, η. As we will see later (Chapter 3, Appendix C), this variable, for light particles, is just single particle longitudinal phase space. Therefore, in the absence of some overall dynamics, we expect particles to be uniformly distributed in pseudorapidity. Since spin and polarization effects are known to be small in

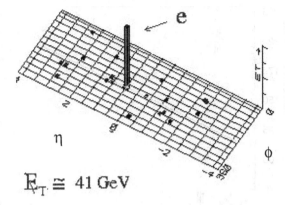

$$\mathbb{E}_T \cong 41 \text{ GeV}$$

Figure 2.7 Schematic display of an event where a single *W* boson is produced and decays into an electron and a neutrino. The "pixels" or calorimetric segments in the plane are defined to be the azimuthal angle and the pseudorapidity. The vertical axis is the transverse energy (CDF – with permission).

proton–(anti)proton collisions, we also expect particles to be produced uniformly in azimuth. The independent elements, or "pixels," are chosen to have roughly constant area in (η, ϕ) space, where θ is the polar angle of the particle in spherical polar coordinates with the beam direction along the z-axis.

$$\eta = -\ln[\tan(\theta/2)]. \tag{2.7}$$

In Fig. 2.7 we show the display of an event obtained in the CDF detector containing a single produced *W* boson, which decays into an electron and a neutrino. The horizontal axes of the plot are azimuthal angle and pseudorapidity and the vertical axis is transverse energy. The "pixels" each give an independent energy measurement. The *W* gauge bosons can decay into quark–antiquark pairs, e.g. $W^+ \rightarrow u + \bar{d}, c + \bar{s}$, or into lepton pairs, $e^+ + \nu_e, \mu^+ + \nu_\mu, \tau^+ + \nu_\tau$. For these two body decays, $E_T \sim M_W/2 \sim 40$ GeV for symmetric decays, as is observed in Fig. 2.7.

Approximately all the energy is deposited in a single "pixel" of the electromagnetic calorimeter. This fact, and the existence of an associated track, with matching momentum, give us electron particle identification. Note also that the existence of a neutrino in the final state is inferred by the failure to balance transverse energy. The missing transverse momentum is indicated by the symbol $\mathbb{E}_T$ and is 41 GeV.

Electromagnetic calorimeters may be calibrated in energy by exposing them to well-prepared particle beams and recording the energy deposit. They may also be calibrated "*in situ.*" In Fig. 2.8 we show the calibration of an electromagnetic calorimeter using the two-photon decay of the neutral pion. The data come from the D0 experiment, which operates at the Fermilab Tevatron collider facility, along with the CDF experiment.

In Fig. 2.9 we show the CDF calibration using the tracker for the charged pion and calorimetry for the neutral pion in $\rho^{+-} \rightarrow \pi^{+-}\pi^0$ decays.

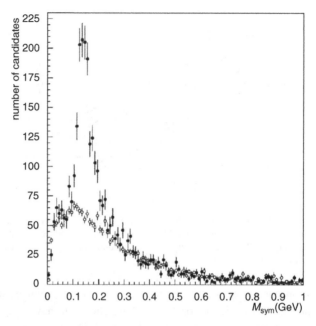

Figure 2.8   Distribution of the invariant mass of two photons in data taken with the D0 calorimeter. Note the resonant peak at the mass of the neutral pion, $M = 0.14$ GeV, and the experimental width. The smooth background curve arises when uncorrelated photons from different events are used ([3], D0 – with permission).

2.4 Hadron calorimetry – jets of q and g and neutrino (missing E_T)

The outer longitudinal compartment of the calorimetry in a general purpose detector serves to detect and measure "hadrons," or strongly interacting particles. We must be careful to define the hadrons, because we have been imprecise so far. We have thus far defined the strong force to be the long range (massless gluons) interaction between colored quarks mediated by colored gluons. However, colored objects appear experimentally to be absolutely confined, e.g. no free quarks are found, so that isolated quarks and gluons do not exist, only the colorless "hadron" combinations of quark–antiquark or three quark bound states.

There are residual forces between these "hadron" states which are responsible for binding protons (*uud* bound state) and neutrons (*ddu* bound state) together in the nucleus. That force is observed to be strong (it overcomes the Coulomb repulsion of the protons in the nucleus) and short ranged. An analogous situation exists in atomic physics. The long range electromagnetic force exists between charged electrons and protons causing neutral atoms to be formed. A residual Van der Waals force between these uncharged atoms is short ranged ($\sim 1/r^6$) and results in the formation of molecules, bound states of neutral atoms. Typically, we will concentrate on the quark and gluon interactions, as the complex hadron interactions are really "quark molecular chemistry" and we aim to study the fundamental interactions. However, in discussing calorimetry we need to explore the hadrons themselves.

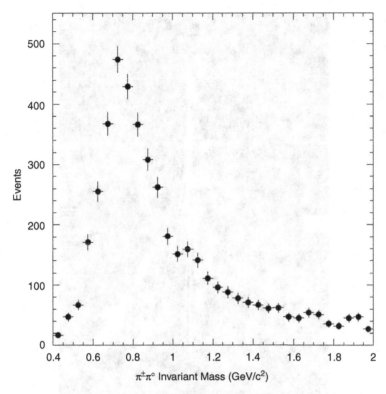

Figure 2.9   Distribution of the invariant mass of two pions in data taken with the CDF calorimeter. Note the resonant peak at the mass of the ρ meson, $M = 0.769$ GeV. ([4], CDF – with permission).

A typical hadronic interaction is shown in Fig. 2.10. Note the limited transverse momentum, or small emission angle, of the secondary particles. Note also the high number of secondary particles produced in a single interaction. The large final state multiplicity is in contrast to electromagnetic processes where there are only two produced particles per incident particle.

There is a characteristic transverse momentum in inelastic hadronic collisions, which is about 0.4 GeV. Crudely speaking, the secondary particles that are produced are all pions, and pions with charge plus, minus and zero are all equally produced. Pions are the lightest hadrons, quark–antiquark bound states ($\pi^+ = u\bar{d}, \pi^\circ = u\bar{u}, d\bar{d}, \pi^- = d\bar{u}$). The neutral pions decay rapidly into two photons, which are then detected as showers in a fashion similar to that discussed above in the section on electromagnetic calorimetry. The charged pions decay weakly, with decay distances much larger than the detectors we describe here, so we can consider them to be stable.

However, the pions do continue to interact. There is a characteristic length over which a hadronic interaction occurs, the interaction length λ_o, which is the mean free path of the pion to suffer a strong interaction. In iron this length scale is 16.8 cm. In order to completely absorb, and hence measure, the energy, a total path length of at least ten

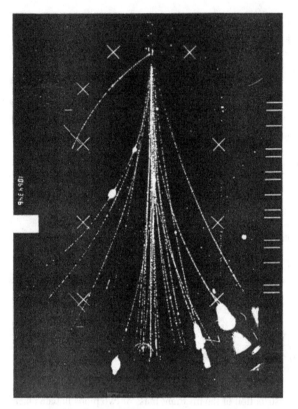

Figure 2.10   Photograph of a 200 GeV pion interaction ([5] – with permission).

interaction lengths is needed, or a calorimetric "depth" of ~1.7 m. In Fig. 2.11 the absorber structure is shown for a typical hadronic calorimeter. The structures are clearly not as compact as those of electromagnetic calorimeters.

In analogy to the critical energy in electromagnetism, there is a threshold energy, E_{Th}, below which new particles cannot be produced. The "threshold" energy for a pion to produce another pion by way of the reaction $\pi + p \rightarrow \pi + \pi + p$ is $E_{\text{Th}} \sim 2m_{\pi} \sim 0.28\,\text{GeV}$. This energy is much larger than the electromagnetic critical energy. Therefore, the number of particles produced in a hadronic shower at "shower maximum", $N \sim E/E_{\text{Th}}$, will always be smaller than the number produced in an electromagnetic shower. Since the energy resolution of a calorimeter is at least partially defined by the stochastic fluctuation in the number of particles in the shower, we also expect that the ultimate energy resolution for hadronic calorimetry will not be as precise as that for electromagnetic calorimetry.

$$dE/E \sim dN/N \sim 1/\sqrt{N} \sim \sqrt{E_{\text{Th}}/E}. \tag{2.8}$$

For example, using Eq. (2.8) to estimate the "stochastic coefficient" in Eq. (2.1), we find $a \sim 53\%$ when E is given in GeV units. That value is, as expected, much larger than the coefficient quoted for electromagnetic calorimetry.

Figure 2.11   Photograph of the absorber of the CMS hadronic calorimeter (HCAL). Note the slots interspersed in the brass absorber structure for the insertion of active detection (sampling) elements. Note also that the total depth of the absorber is about 1 m (Fermilab – with permission).

Sometimes the hadronic or electromagnetic compartment is itself longitudinally segmented. In Fig. 2.12 we show the energy deposit in an initial seven absorption length compartment vs. the energy deposit in the subsequent four absorption lengths. In some cases substantial energy is deposited in the rear compartment. This implies that, were the calorimeter truncated so as not to include the back compartment, the energy resolution would be seriously degraded by fluctuations in the longitudinal shower development and subsequent fluctuations in the energy loss due to leakage out of the back of the calorimeter.

There is, however, an intrinsic limit to the depth. It makes no sense to construct a device that is very thick because an emitted gluon can virtually "decay," or split into a heavy quark, Q, pair, $g \rightarrow Q + \overline{Q}$, with a probability $\sim \alpha_s/\pi$. Subsequent decays of the type $Q \rightarrow q + e^- + \overline{\nu}_e$ occur with a branching ratio $\sim 10\%$. Therefore, a gluon jet will "leak" $\sim 1/6$ of its energy due to escaping neutrinos roughly 0.3% of the time.

The calorimeter shown in the photo of Fig. 2.11 is of the sampling variety. Active detection elements are inserted in the slots that are interspersed in the absorber. An example of a possible active element is shown in Fig. 2.13. In this case optically independent "scintillating" tiles are read out by "wavelength shifting" optical fibers. This type of layout allows us to produce a hadronic calorimeter that has active samples covering almost all the solid angle. A "hermetic" construction is needed if the missing energy is to be accurately measured. Clearly, "dead" regions in the calorimetry are to be avoided since particles lost in them would mimic the emission of undetected neutrinos.

All calorimeter detection elements must be manufactured to achieve a good uniformity. Otherwise, variations of the shower locations in depth or in different "pixels" will lead to variations in the reported energy for a monoenergetic incident particle. For example,

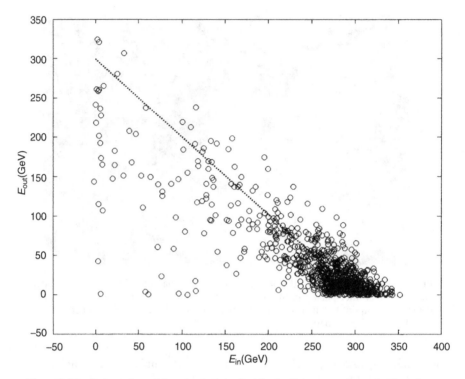

Figure 2.12 Scatter plot of the energy deposited in the first seven absorption lengths of the CMS hadronic calorimeter (x axis) vs. the deposited energy in the next four absorption lengths (y axis). The dotted line indicates a total of 300 GeV deposited in summing both compartments.

in a hadronic calorimeter a variation in light output of the tiles shown in Fig. 2.13 with a standard deviation of 10% leads to a fractional energy error (the factor b in Eq. (2.1)) of about 3%. Similar, but much more exacting, uniformity is needed for the high precision electromagnetic calorimetry.

Calorimeters are often calibrated using prepared beams at accelerators with well-defined momentum. In addition, we can use cosmic ray muons since they deposit a well-defined energy (minimum ionizing particle) in each tile. As we mentioned above, a muon traversing the sampling layers of a calorimeter will deposit only ionization energy. In Fig. 2.14 we show the output signal due to the passage of a muon. This peak is well resolved from the "pedestal" peak that corresponds to zero energy deposit, broadened by noise in the electronics readout. Clearly, calorimetry can also be used in muon "particle identification."

What about the required extent of angular coverage? We know that we want to detect all particles that are emitted in an event, so as to infer the vector momentum of an emitted, and undetected, neutrino. However, technically we cannot achieve total coverage due to the necessary existence of vacuum pipes containing the proton–(anti)proton beams and the obstacles due to the magnetic focusing elements of the accelerator, for example.

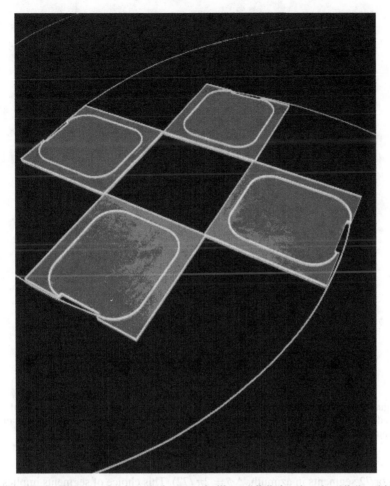

Figure 2.13   Photograph of a calorimeter scintillator "tile" showing the "tile" and its "wavelength shifting" fiber. The optical signal is converted from blue light in the "tile" to green light in the fiber and then captured and taken out through the small fiber (Fermilab – with permission).

How small an angle do we need to cover? Figure 2.15 shows the pseudorapidity distribution of particles that we wish to detect after they emit a virtual W or Z gauge boson, for example, by way of a "radiative" process, where a d quark bound into the initial state proton radiates a W^- and turns into a u quark, $d \to u + W^-$. These processes are very important in Higgs searches, so that calorimetry should extend to $|\eta| \sim 5$, or to a polar angle of about $0.8°$, at the LHC experiments.

Previously, in Fig. 2.7, we saw the electron signal in an event where a produced W boson decayed into an electron and a neutrino. The calorimetry information was shown as the transverse energy deposited in independent (η, ϕ) "pixels." What sort of angular size is needed? In Fig. 2.16 we see a choice with pixel widths $\Delta\eta \sim \Delta\phi \sim 0.087$ (η is dimensionless and the units for ϕ are radians so that, in this example, there are five degree

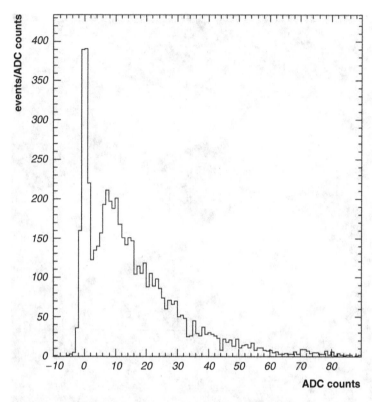

Figure 2.14   Distribution of the deposited energy in a calorimeter tile. Note the "pedestal" due to zero energy deposit and the ionization peak due to the passage of a muon ([5], CMS – with permission).

pixels, or 72 segments in azimuth, $\Delta\phi = 2\pi/72$). This choice of segments implies that we can resolve a Higgs of 1 TeV mass decaying into ZZ that in turn decay into four quarks.

A 1 TeV Higgs decays at rest into a ZZ pair, each with a momentum ~500 GeV. The subsequent decay of a Z into quark pairs, for massless quarks, has a total transverse momentum between the quark and antiquark equal to the Z mass or ~91 GeV for symmetric decays. The opening angle between the quarks is ~0.2 radians. These quarks then go into separate calorimetric segments of full width 0.087 and can be resolved as two distinct objects (see discussion in Chapter 5). Since there is a theoretical upper limit on the Higgs mass of roughly 1 TeV, this choice of pixel size for HCAL is acceptable.

We have so far discussed hadrons and evaded the question of how we detect quarks and gluons. These latter objects have color, and color is thought to be completely confined. We assert that the color force is weak at small distances and strong at large distances (see Appendix D). As a result colored objects cannot be separated beyond a distance set by a QCD parameter that has a characteristic length ~1 fm ~ $1/\Lambda_{QCD}$. Therefore, the quarks or gluons must shed their color by becoming an ensemble of colorless hadrons, for

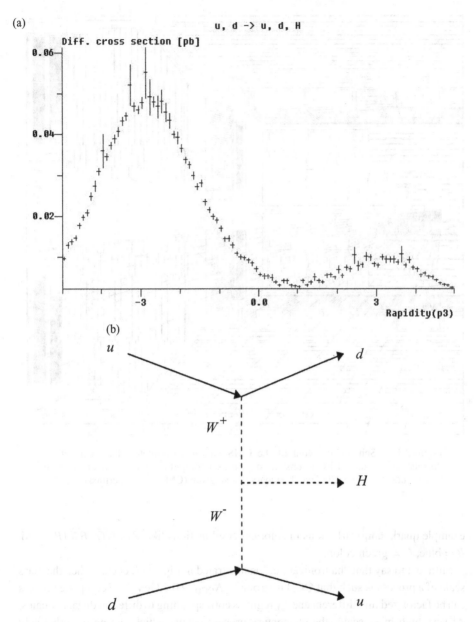

Figure 2.15   (a) Distribution in pseudorapidity of the recoil or "tag" jets produced in the *WW* fusion process of Higgs production. (b) Feynman diagram for the *WW* fusion process $u + d \rightarrow d + u + H$.

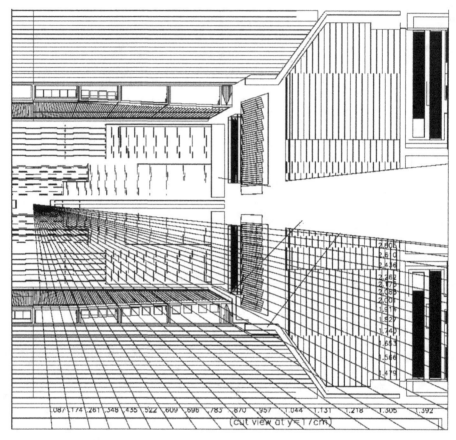

Figure 2.16 Schematic layout of the CMS hadron calorimeter. The segments or "pixels" are separated by a constant step in pseudorapidity and in azimuthal angle. The pseudorapidity of the pixel boundary is also given (CMS – with permission).

example quark–antiquark pions in colorless combinations like $R\overline{R}$, $G\overline{G}$, $B\overline{B}$ (R = red, B = blue, G = green colors).

Suffice it to say that "hadronization," as illustrated in Fig. 2.17, occurs when the mass scale of a process is such that QCD is strong, $\sim \Lambda_{QCD} \sim 0.2$ GeV. The complete process can be factorized into different energy regimes corresponding to different distance scales. At very high mass scales the elementary process occurs which can be perturbatively calculated because the color interaction is weak. At moderate masses, or transverse momentum scales, $P_T \gg \Lambda_{QCD}$, perturbative QCD can still be used, and the colored quarks and gluons radiate in a QCD "shower."

When the strong interactions become strong, these colored objects become "bleached" and evolve into an ensemble, or "jet" of colorless hadrons. The quark or gluon "jet" is expected to look something like Fig. 2.10. The "jet" of hadrons that emerges has the approximate direction and momentum of the parent quark or gluon.

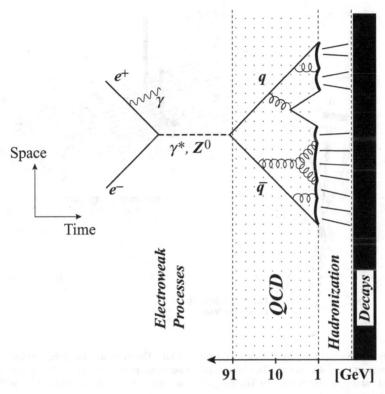

Figure 2.17 Schematic representation of the evolution of quarks produced in the final state of an interaction. In the high energy regime the quarks are almost free particles and the process is calculable in perturbation theory. In the intermediate energy range we can again use perturbative QCD. At an energy range where the scale factor for QCD is the typical energy, 0.2 GeV, hadronization and strong decays of hadron resonances occur, which must be treated phenomenologically because the coupling is strong ([7] – with permission).

Unstable particles like the W, Z, and top quark all have decay widths ~ 1 GeV. Therefore they decay in a distance 0.2 fm, before they "hadronize" at a distance scale of ~ 1 fm. This is why there is no "toponium" – the QCD bound state of a top and antitop quark. It decays, $t \rightarrow b + W^+$, before the bound state can form.

The scattering of the quarks that we, for now, simply assume to exist inside the proton leads to a "jet" of particles traveling in the direction of, and taking the momentum of, the parent quark. We assume that the proton and (anti)proton contain quarks and gluons, which have a limited transverse momentum $\sim \Lambda_{QCD}$. A "dijet," or two jet, event is shown in Fig. 2.18. There is energy in both the electromagnetic (lower) and hadronic (higher) compartments now, as opposed to the case shown in Fig. 2.7, when the electron deposited all its energy in the electromagnetic compartment. Note also that the "jets" are spread over several pixels. The two jets are, however, reasonably well collimated and are approximately "back-to-back" in azimuthal angle, $\phi_1 - \phi_2 \sim \pi$.

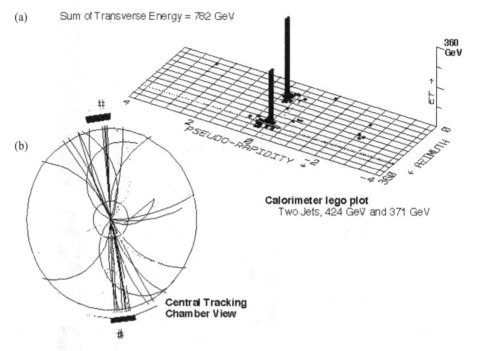

(a)   Sum of Transverse Energy = 782 GeV

(b)

Calorimeter lego plot
Two Jets, 424 GeV and 371 GeV

Central Tracking Chamber View

Figure 2.18   (a) Display of a two jet event at CDF. The vertical axis is the transverse energy in the calorimeters. The horizontal plane consists of the "pixels" in azimuthal angle and pseudorapidity. (b) Tracking detector data for the dijet event (CDF – with permission).

The tracking detector azimuthal–radial plot for that event is also shown in Fig. 2.18. Recall that large momentum corresponds to small "bend" angle in the magnetic field. Clearly the jet has an internal particle structure. There is a "core" of fairly high momentum particles near to the axis of the jet, with lower momentum particles associated with the jet but emitted at larger angles to the jet axis. The magnetic field also has the effect of "sweeping" the lower momentum particles away from the jet axis, as can be seen in Fig. 2.18(b).

A polar angle projection of a D0 "dijet" event is shown in Fig 2.19. Again the jets are fairly well collimated in solid angle, deposit energy in both compartments of the calorimeter, and are also roughly back-to-back in polar angle.

The description of "hadronization" necessarily has recourse to experimental data on the momentum distribution of hadrons found in jets (e.g. Fig. 2.18). Representative data are shown in Fig. 2.20. We simply define a distribution of the hadronic "fragments" of the quark or gluon in z, $D(z)$, where z is defined to be the fractional jet momentum taken off by the hadronic fragment, $z = P_{\text{hadron}}/P_{\text{jet}}$. The distribution $D(z)$ is roughly of the form $zD(z) = (1 - z)^a$.

The efficiency to "tag" a jet as having originated from a heavy flavor parent (e.g. a b quark) depends on the momentum of the jet. Higher energy jets have longer decay

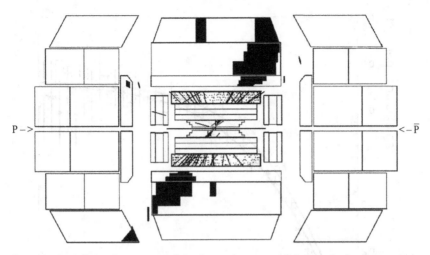

Figure 2.19   Schematic representation of a two jet event at D0. The shading represents the scale of energy deposited in the calorimeters. The first compartment is the electro-magnetic calorimeter followed by two hadronic compartments. This is a projection in polar angle ([8] – D0 – with permission).

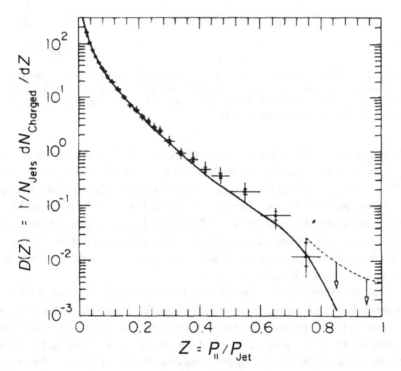

Figure 2.20   Distribution $D(z)$ from CDF data of the fractional energy of a hadronic jet fragment. Note the steep falloff with increasing z ([9] – with permission). The lines are fits to the data points.

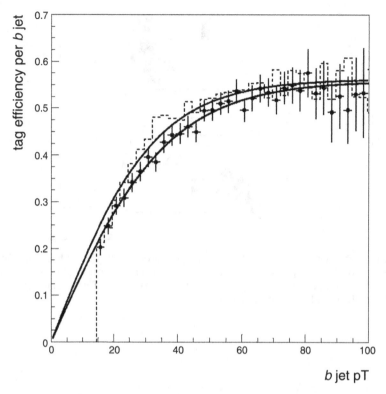

Figure 2.21　Efficiency for tagging a heavy flavor jet as a function of the transverse momentum of the jet. The efficiency for false tagging of light quark or gluon jets is small under these conditions at all momenta ([10] – with permission). The lines are the results of different Monte Carlo models.

lengths (relativistic time dilation). However, the existence of many confusing fragments at high energies means that the ability to find the secondary vertices is not perfect. Therefore, if we wish to suppress the large background of light (u, d) quark or gluon jets to an acceptable level, the efficiency to tag the b jet will be reduced. Multiple scattering error makes the rejection more difficult for a low jet momentum. Monte Carlo predictions from the CDF experiment at the Fermilab Tevatron are shown in Fig. 2.21. Note the rise of the efficiency with transverse momentum to a level of ∼50% for jet transverse momentum >50 GeV.

Detection and measurement of jets is accomplished by way of calorimetric determination of the energy of a localized ensemble of hadrons. We need to know how accurately we can measure the jets given that we know the single particle resolution of a calorimeter. We assume we know the numerical value of the constants a and b in Eq. (2.1). Single particle data are available by utilizing test beams supplied at accelerators, for example.

The ensemble energy is the sum of the single particle energies, $E = E_1 + E_2 + E_3 + \cdots$. If the stochastic term dominates in the error on the measurement of individual hadrons, we find that the energy resolution of the ensemble is the same as the single particle resolution, $dE/E \sim a/\sqrt{E}$. Therefore, if we have typical values like a ∼50%

(GeV)$^{1/2}$ and $b = 3\%$, we expect to measure the energy of a 100 GeV quark or gluon jet with an accuracy of $\sim 5\%$.

$$dE^2 = dE_1^2 + dE_2^2 + dE_3^2 + \cdots$$
$$\approx aE_1 + aE_2 + aE_3 + \cdots = aE. \tag{2.9}$$

In the very high energy case where only the constant term is important, the ensemble is measured more accurately than the single particle.

$$z_i \equiv E_i/E,$$
$$dE/E = b\sqrt{z_1^2 + z_2^2 + z_3^2 + \cdots}. \tag{2.10}$$

If the energies of the jet fragments are equi-partitioned, there are n terms of equal magnitude $z_i = 1/n$ and the series in Eq. (2.10) can be summed:

$$dE/E \sim b/\sqrt{n}. \tag{2.11}$$

For a jet–jet mass, M, measurement, we assume that the angular error is not the dominant error. This will be the case for objects whose momentum is less than their mass because then the angle between the jets is large and, hence, insensitive to small errors. Note that the two body mass is $M^2 = (P_1 + P_2)_\mu \cdot (P_1 + P_2)^\mu \sim 2P_{1\mu} \cdot P_2^\mu = 2(E_1 E_2 - \vec{P}_1 \cdot \vec{P}_2)$. For massless jets "decaying" approximately at rest the error on the dijet mass due to the energy errors on the two jets can be calculated assuming $\cos \theta_{12} \sim -1$, $E_1 \sim E_2 \sim E \sim M/2$. For a 100 GeV mass, a 5% measurement is expected.

$$M^2 = 2E_1 E_2(1 - \cos \theta_{12}) \approx 4E_1 E_2 \sim 4E^2,$$
$$dM/M \sim a/\sqrt{2E} \sim a/\sqrt{M}. \tag{2.12}$$

The reconstructed mass of a W boson decaying into two quark jets is shown in Fig. 2.22. The W mass is measured with a standard deviation of ~ 3 GeV. Thus the fractional mass error is $\sim 3.75\%$, which is of the expected order of magnitude. In addition, precise energy information on individual hadrons from the tracking subsystem can also be used for the charged hadrons. This technique will allow us to improve the kinematic measurements of the jets beyond the accuracy made available by purely calorimetric methods.

The neutrinos are "measured" indirectly by looking at the "missing transverse energy," assuming that the initial state has zero transverse energy. This measurement involves a "collective" variable, as all the transverse energy in an interaction must be measured in order to find out how much is missing. There are errors due to the limited angular coverage of the detectors, the finite energy resolution of the calorimeters, and the failure of low momentum particles to even reach the calorimeters if there is a strong solenoid magnetic field.

In the idealized case of an interaction containing only two jets with no longitudinal momentum, the jet energies are $E_1 \sim E_2 \sim M/2$. We assume that the stochastic term dominates the energy resolution. The missing transverse energy is denoted by $\not{E}_T$. The missing transverse energy due to simple jet energy mis-measurement is then $\not{E}_T \sim E_1 - E_2$.

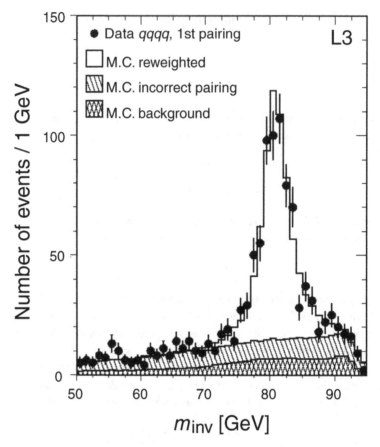

Figure 2.22 Distribution of the dijet mass reconstructed from energy measured in a calorimeter. Note the resonant peak at the W mass and the experimental width due in part to the errors in the energy measurements ([11] – with permission).

The error on the missing energy is:

$$dE_T \sim a\sqrt{M} = dM \sim a\sqrt{\sum E_T}. \tag{2.13}$$

Therefore an event containing a dijet of mass 100 GeV has a total transverse momentum of $\sim$5 GeV due to jet energy mis-measurement, if $a \sim 50\%$ (see Fig. 2.18). We assert that the approximate generalization to the case of many jets in the final state is as shown in Eq. (2.13), where we sum over the transverse momentum of all particles produced in the interaction.

We now have approximate expressions for the expected calorimetric error for jet energy, dijet mass, and missing transverse momentum. We will use these estimates in our discussions of search strategies for the Higgs boson.

An event with a single W boson produced that decays into an electron and a neutrino is shown for the D0 detector in Fig. 2.23. The electron energy goes entirely into the

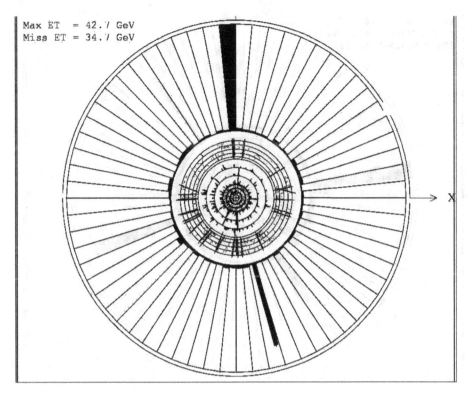

Figure 2.23   Schematic azimuthal–radial view of a D0 event with a single *W* in the final state. The missing energy in the event is close to being back-to-back with the deposited electron energy (D0 – with permission).

electromagnetic compartment (in the +*y* direction here). The missing energy measured in the calorimetry, electromagnetic and hadronic, is also shown (in ∼ the −*y* direction), indicating the two-body nature of the *W* decay.

Another event with missing energy in the final state is shown in Fig. 2.24. In this case a *W* and *Z* boson are produced, where the *W* decays into *e* + *v*, while the Z decays into an $e^+ + e^-$ pair. Note the back-to-back nature of both the Z and *W* decays, indicating that the *W* and the Z are both produced with little transverse energy, and that the missing energy roughly balances the transverse energy of the electron from the *W* decay.

Transverse momentum balance can also be used as a constraint for *in situ* detector calibration. The transverse energies are simply assumed to balance on average, and this assumption is used to extend the calibration of the mean from a calibrated pixel to an uncalibrated one. The procedure is illustrated in Fig. 2.25.

2.5 Muon systems

The muons exiting from the vertex are charged particles, and have their vector position and momentum measured accurately first in the tracking subsystem. However, muons

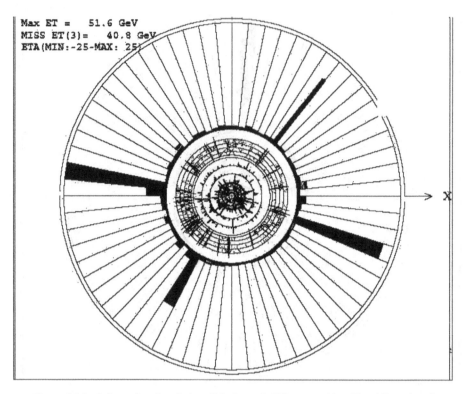

Figure 2.24   Schematic azimuthal–radial view of a D0 event with a W and Z produced in the final state. The W decays into an electron ($\sim -y$) and a neutrino ($\sim +y$) while the Z decays into an electron–positron pair ($\sim +x$ and $-x$) ([12] – D0 – with permission).

are rarely produced, and our job is to pick out which track is a muon in order to trigger on it. Triggered events pass the first selection of those rare events which will be part of the data logging. Particle identification is achieved by exploiting the fact that muons (of energy <300 GeV) do not radiate appreciably, nor do they have strong interactions. Therefore, they pass through the calorimetry, depositing only ionization energy (see Fig. 2.13). As they pass through the return yoke of the magnet, all the other particles have been absorbed by the calorimetry, see Fig. 2.11.

Therefore, particles that are detected in the muon system are assumed to be muons, and the issue is to trigger cleanly on these rarely produced particles. The most accurate momentum measurement of the muon comes from the tracking subsystem, while a redundant momentum crosscheck and particle identification comes from the muon tracking chambers. The two distinct measurements are illustrated schematically in Fig. 2.26.

The main functions of the muon system are to perform particle identification on the muons and to provide a muon trigger. The trigger is drastically simplified because almost the only particles that survive to enter the muon detectors are muons. Therefore, the first

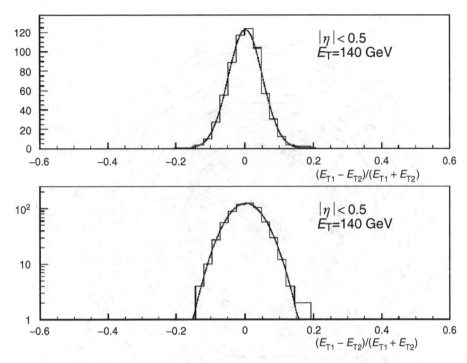

Figure 2.25 Distribution of the fractional jet transverse energy difference in dijet events. Note the sharp peak at 0 and the steep falloff of the distribution, which appears to be almost a pure Gaussian. The line is a Gaussian fit to the data. ([8] – D0 – with permission).

task is to "pattern recognize" a clean trajectory in the muon detectors in an environment which is quite sparsely populated.

The trigger also requires a reasonably accurate measurement of the muon transverse momentum. A good measurement is needed because there are many low transverse momentum muons, which are of little interest. These muons arise from heavy quark, Q, decays, $Q \to q + \mu^- + \bar{\nu}_\mu$, where the Q may itself arise from a virtual gluon "decay," $g \to Q\overline{Q}$. These muons are copiously produced (Fig. 2.27) and must be rejected in the trigger lest they swamp the higher momentum muons of interest that are due to the decays of W and Z bosons and other rarely produced objects.

The task becomes clear when we explore the source of muons at proton–(anti)proton colliders. Muons from the produced b and c quarks dominate at low transverse momentum, where the scale is set by the b quark mass, ~ 5 GeV. At higher momenta, where the scale is given as one half the gauge boson W or Z mass (two body decay) or ~ 40–45 GeV, the main source of muons is the decay of gauge bosons (see Table 2.1). There are no mass scales yet known above this, so searches for new heavy particles are made in the tails of the distributions of muons from W and Z decay.

The invariant mass distribution of dimuon events from D0 is shown in Fig. 2.28. The two body decay $\psi -> \mu^+ \mu^-$ with a ψ resonant mass of ~ 3.1 GeV is observed. The

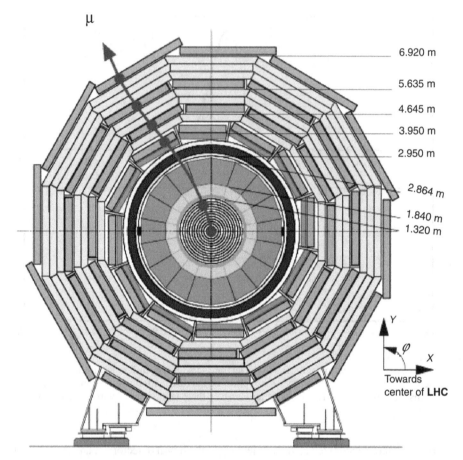

Figure 2.26   Schematic azimuthal-radial layout of muon detection in the CMS experiment. The muons are first bent in the central magnetic field and detected/measured
in the tracking subsystem. After traversing the calorimetry and magnet coil the muon
is subsequently bent in the steel return yoke and re-measured in the muon chambers
embedded in the steel ([1], CMS – with permission).

ψ is a narrow bound state of a charm–anticharm quark pair. This resonant peak can be
used to check the calibration and alignment of the muon chambers *in situ*.

The mass resolution shown here is rather poor. This is because the momentum used
in this plot was determined solely from the muon chambers. Since the chambers are
interspersed in an iron return yoke, the momentum measurement is limited by multiple
scattering (see Eq. (2.1)) to a $\sim$15% error. The momentum impulse, or change in transverse momentum, due to the magnetic field B existing over a distance L is $\sim BL$. The
multiple scattering impulse in traversing that same region is $\sim \sqrt{L}$, where the square root
is characteristic of stochastic behavior. Thus the ratio, which determines the fractional
momentum resolution, scales as $\sim 1/B\sqrt{L}$. The magnetic field is limited by iron satu-
ration to $\sim$2 T. The length of steel is limited by financial and mechanical considerations

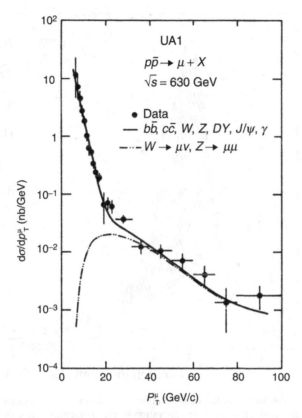

Figure 2.27 Distribution of the transverse momentum of muons measured in the UA1 collider experiment at CERN. The two main sources of muons are the decay of heavy quarks at low transverse momentum and the decay of W and Z gauge bosons at high transverse momentum ([13], UA1 – with permission).

to ~ 1 m, hence, the limited momentum resolution for muons measured in steel. The multiple scattering impulse is $(\Delta P_T)_{MS}$, while the magnetic field impulse is $(\Delta P_T)_B$.

$$dP/P \sim (\Delta P_T)_{MS}/(\Delta P_T)_B \sim 0.15. \tag{2.14}$$

In order to obtain a better measurement, we would have to supply tracking chambers in a volume with magnetic field and without multiple scattering. If this is done after the calorimetry the tracking is clean because almost only muons survive. However, it makes for a large, and hence expensive, detector. If we instead wish to use the inner tracking system, we must extrapolate the track from the muon system back into the tracking chambers and attempt to match tracks in vector position and momentum. This matching procedure is, in turn, limited by the multiple scattering errors induced by passage of the muons through the calorimetry that separates the two tracking systems. Different experiments have made different choices. There is no "correct" decision in this matter in any event.

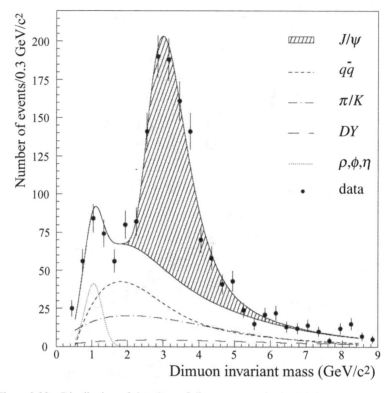

Figure 2.28   Distribution of the mass of dimuon events in the D0 detector. Note the resonant ψ peak, which is used for calibration of the momentum scale of the muon detectors. Note also that the width is set by the multiple scattering of the muons in the steel and is not due to the intrinsic accuracy of the chambers ([14], D0 – with permission). The lines indicate background sources of dimuon events.

2.6 Typical inelastic events

The vast majority of interactions in a proton–(anti)proton collider are uninteresting. They occur at low mass scales $\sim \Lambda_{QCD}$ where the dynamics is strong, and hence difficult to compute. The secondary particles in such a collision have low transverse momentum, $P_T \sim \Lambda_{QCD}$. We are interested in high mass states, which implies final state particles with a large transverse momentum.

Many of the interesting physics processes that we will discuss in the later chapters have pb (1 pb $= 10^{-36}$ cm^2) cross sections, while the total inelastic cross section, making "minimum bias," or inclusive inelastic events, is ~ 100 mb, which is 100 billion times larger. Obviously, we are looking for rare processes and we need to trigger incisively, as noted previously.

It also must be remembered that, even though we have an "interesting" process occurring at large P_T in an interaction, there are also all the soft fragments of the remaining quarks and gluons that hadronize and form the "underlying event." Indeed, most of the particles in an "interesting" event are themselves uninteresting. Furthermore, the

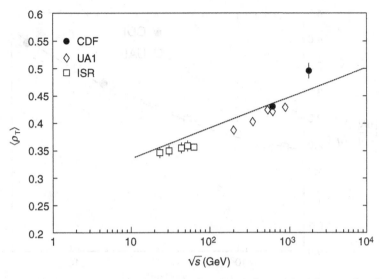

Figure 2.29   Mean transverse momentum of produced charged particles as a function
of the center of mass energy in *p*–(anti)*p* collisions. Note the logarithmic dependence
on CM energy indicated by the line ([15], CDF – with permission).

detectors we use may not be fast enough to resolve individual interactions. In that case
we have a "pileup" of "minimum bias" events within the resolving time of the detector.
We need to understand some of the basic features of these events because they form
an irreducible background on top of which resides the interesting high P_T fundamental
interaction wherein new discoveries may lie.

In Fig. 2.29 we display a plot of the mean transverse momentum of all produced
charged particles in "minimum bias" events or typical inelastic interactions. This quantity,
$\langle P_T \rangle$, is a weak function of the total available CM energy. At 10 TeV, it is roughly
0.5 GeV.

The scale for the mean transverse momentum is the QCD scale, which is not unex-
pected.

$$\langle P_T \rangle \sim \Lambda_{\text{QCD}}. \qquad (2.15)$$

We assert that π^+, π^0, and π^- are produced in roughly equal numbers and are the dominant type of hadrons produced in inelastic collisions. Pions are produced approximately
uniformly in pseudorapidity. The density of charged particles per unit of pseudorapidity
is shown in Fig. 2.30. It is a weak function of the CM energy. At 10 TeV the density is
extrapolated to be about six charged particles per unit of pseudorapidity, or about nine
pions per unit of η.

Therefore, each "minimum bias" interaction measured in a detector that operates at
the 14 TeV CM energy of the LHC and fully covers angles with $|\eta| < 5$, creates 90 =
$10 \times (6 + 3)$ charged and neutral pions with a total scalar transverse energy deposit of
45 GeV. We assert that the "underlying event" in a "hard" or high transverse momentum
collision is approximately the same as a "minimum bias" event.

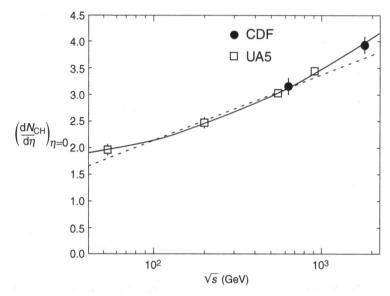

Figure 2.30 Mean number of produced charged particles per unit of pseudorapidity as a function of CM energy in p–(anti)p collisions. Note the rough logarithmic dependence of particle density on center of mass energy ([16], CDF – with permission). The lines represent fits to the data.

If we are operating at high interaction rates, such as are expected at the Large Hadron Collider (LHC) at CERN, there may be 20 "minimum bias events" in a beam–beam bunch crossing that cannot be temporally resolved. This is the minimum "pileup" because two bunched beam crossings are separated by only 25 nsec and we need to use very fast detectors if we are to have "only" 20 overlapping events. The minimum "pileup" is then a beam bunch containing 1800 particles with 900 GeV of deposited transverse energy. If we blindly apply Eq. (2.13), we expect ~15 GeV of missing transverse energy, on average, in each bunch crossing, simply due to the calorimetric energy error made in measuring all the particles in the bunch crossing.

A jet is typically defined to be an ensemble of particles possessing a large transverse energy deposited in a small circular region of radius R, in (η, ϕ) phase space, $R < 0.7$. A finite jet size in R is required if we are to record all the jet energy, as seen in Figs. 2.17 and 2.18. Since there is a substantial "pileup" of transverse energy, false jets may be detected at low jet transverse energies ~30 GeV, while at higher jet energies the extra pileup energy must also be accounted for and the jet energy corrected.

Triggers and reconstruction algorithms can look at the transverse energy flow within the jet cone to select real jets, which have a "core" as opposed to approximately uniformly distributed pileup. For example, a cone of radius $R \sim 0.7$ contains more than one hundred pixels of the size shown in Fig. 2.16. That granularity is sufficient to resolve the details of energy flow within the cone defining the total jet energy. Jets have a limited momentum transverse to the parent direction, k_T, and a distribution $D(z)$ of the momentum of the hadronic fragments with a "leading" hadronic fragment taking off, on average, a fraction, $\langle z_{max} \rangle \sim 0.2$, of the parent jet energy – see Fig. 2.20.

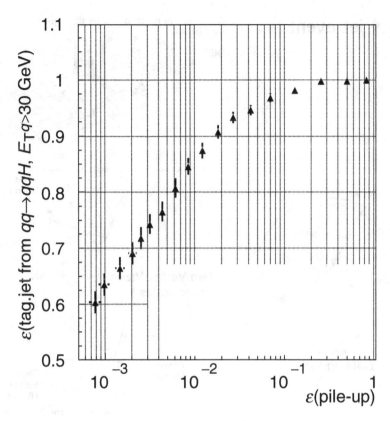

Figure 2.31   Efficiency for the rejection of fake jets with respect to the efficiency for finding tag jets at design luminosity in the CMS detector. A cut is first made on the leading pixel transverse energy for events with a cone energy of 30 GeV (CMS – with permission).

The "pileup" transverse energy found on average in any "cone" of radius $R \sim 0.7$ is 20 events $\times$ 0.5 GeV/particle $\times$ 9 particles/area $\times$ $(\pi R^2)/2\pi \sim 22$ GeV. We must use the additional information on the structure of the energy flow within the jet to reduce the number of false jets due to pileup.

As seen in Fig. 2.31, a cut on the transverse energy flow within a cone is a good discriminant between jets with transverse energy around 30 GeV and "fake jets." The signal in Fig. 2.31 consists of "tag jets" from the WW fusion process (Fig. 2.14), while the background is due to pileup of $<n> = 17.3$ minimum bias events, on average. Clearly, asking for a "leading jet fragment" with a large fraction of the total jet transverse energy works fairly well.

2.7 Complex event topologies in D0 and CDF

Clearly, several different fundamental particles of the SM can occur in a complicated event. An example is the CDF event shown in Fig. 2.32. The CDF detector has three main

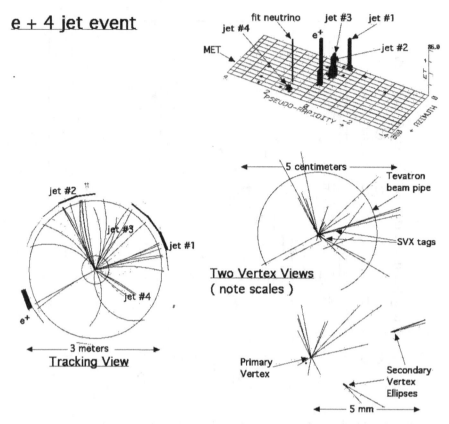

e + 4 jet event

Figure 2.32 A complicated event in the CDF detector. This event contains an electron, four jets, and missing energy due to neutrinos. Note also that there are secondary vertices in the event indicating that some of the jets are the decay products of heavy quarks (CDF – with permission).

detector systems: tracking – silicon + ionization in a magnetic field, scintillator sampling calorimetry, (EM – e, γ and HAD – h), and ionization tracking for muon measurements.

This event contains four jets, as recognized by identifying localized energy deposits in the calorimeter pixels. In addition there is an electron, recognized as energy deposit in the electromagnetic compartment of the calorimetry, with a matching charged track in the tracking detector. There is also a neutrino, as identified by the existence of missing transverse energy in the calorimetry. In addition, two of the jets have secondary vertices in the tracking subsystem, which makes them possible b quark candidates.

A complex event from D0 is shown in Fig. 2.33. The D0 detector has three main detector systems: ionization tracking, liquid argon calorimetry (EM, e, and HAD jets), and magnetized steel + ionization tracking muon, μ, detection/identification.

This event, shown in a polar view, has jets in both the compartments of the calorimetry. It also has a muon candidate ($\sim+y$), which is confirmed by the presence of small ionization energy in the calorimetry and an associated track. In addition, there is an

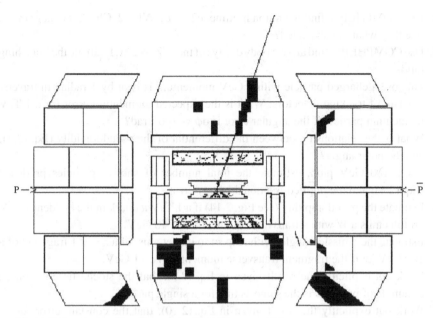

Figure 2.33   A complicated event in the D0 detector. This event contains jets, a muon, an electron, and missing energy (D0 – with permission).

electron candidate with energy deposit only in the electromagnetic compartment (small radius) with an associated track ($\sim -y$). Finally, there is a neutrino candidate in the event, inferred from the missing transverse energy.

The examples given here indicate the complexity of the events that can be studied in general purpose detectors. We conclude that a well-designed general purpose detector can use specialized subsystems to identify and measure photons, electrons, muons, jets of quarks, gluons, and neutrinos. Heavy quarks and leptons are further identified by searching in the tracker for separated secondary vertices. The W and Z gauge bosons decay rapidly and are identified from their decay products.

Particle tracking affords very accurate measurements of electrons and muons. Precision electromagnetic calorimetry provides energy measurements of order 1% (100 GeV energy) for photons and electrons. Gluon and quark jets are measured somewhat more poorly in the hadronic calorimetry, perhaps at the 5% level (100 GeV energy). Neutrinos are also "measured" in the calorimetry, to a similar precision, but the longitudinal component of the neutrino momentum is not well measured due to the necessarily incomplete polar angle coverage of the detectors.

Exercises

1. How far, on average, will a b quark with lifetime $c\tau = 475$ μm and energy 60 GeV travel before decaying?
2. Evaluate the estimated muon lifetime (Eq. (2.2)) with muon mass 0.105 GeV.

3. Use COMPHEP to find the muon lifetime, $e2 \rightarrow e1, N1, n2$. Check the diagram(s). Are they what you expected?

4. Use COMPHEP to find all two body decays of the Z, $Z \rightarrow 2*x$. Evaluate the branching ratios.

5. Suppose a charged particle with 1 GeV momentum is bent by 1 radian in traversing 1 m of tracking detectors. What is the expected momentum error for a 1 TeV momentum particle if the angular error is $d\phi \sim 100$ μrad?

6. What is the relationship between the differential of the pseudorapidity (Eq. (2.7)) and the polar angle?

7. For a 100 GeV pion, estimate the total number of shower particles produced (Eq. (2.8)) and the implied fractional energy error.

8. Estimate the pseudorapidity (see Fig. 2.14) if a 1 TeV u quark in the incident 7 TeV proton emits a W with a transverse momentum of 40 GeV.

9. Estimate the emission angle, with respect to the jet axis, of a $z = 0.1$ fragment of a 100 GeV jet if the fragment transverse momentum is $\sim$1 GeV.

10. Work out explicitly the result given in Eq. (2.9), that the stochastic error on an ensemble of particles is the same as that for a single particle.

11. Work out explicitly the result given in Eq. (2.10), that the constant error for an ensemble of particles is less than that for a single particle.

12. What is the dimuon opening angle for a 10 GeV ψ decay (mass 3.1 GeV)?

13. Use COMPHEP to find the decay width of the τ. Compare with the width quoted in Chapter 2, $e3 \rightarrow n3, 2*x$. Evaluate the six sub-processes to find branching fraction and total width.

14. Use COMPHEP to find the total decay width for the heavy quarks and leptons discussed in this chapter, $e3 \rightarrow 3* x$, $c \rightarrow 3*x$, $b \rightarrow 3*x$. Compare with the data plotted in this chapter.

15. Explicitly work out the threshold for pion production in pion–proton interactions. The energy threshold occurs when the reaction uses all the energy to produce mass and none to give the reaction products kinetic energy. Thus, all particles are at rest in the CM at threshold.

16. Use COMPHEP to look at "tag jets" in $d, u \rightarrow u, d, H$. Plot the distribution of u rapidity and compare with the result given in the text.

References

1. CMS Technical Proposal, CERN/LHCC 94–38 (1994).
2. Leighton, R. B., *Principles of Modern Physics*, New York, McGraw-Hill Book Co. Inc. (1959).
3. D0 Collaboration, *Phys. Rev. D*, **58**, 12002 (1998).
4. CDF Collaboration, Fermilab Pub-01/390-E (2002).
5. Kleinknecht, K., *Detectors for Particle Radiation*, Cambridge, Cambridge University Press (1987).
6. CMS HCAL Group, *Nucl. Inst. Meth. A*, **457**, 75 (2001).
7. Bethke, S., MPI-PhE/2000–02.
8. Montgomery, H., Fermilab–Conf-98-398 (1998).
9. CDF Collaboration, *Phys. Rev. Lett.*, **65**, 968 (1990).
10. Carena, M., Conway, J., Haber, H., and Dobbs, J., arXiv:hep-ph/0010338 (2000).

11. Glenzinski, D. and Heintz, U., arXiv:hep-ex/0007033 (2000).
12. Montgomery, H., Fermilab-Conf-99/056-E (1999).
13. Alterelli, G. and Di Lella, G., *Proton–Proton Collider Physics*, Singapore, World Scientific (1989).
14. D0 Collaboration, *Phys. Rev. Lett.*, **82**, 35 (1999).
15. CDF Collaboration, *Phys. Rev. Lett.*, **61**, 1819 (1988).
16. CDF Collaboration, *Phys. Rev. D.*, **41**, 2330 (1990).

Further reading

Anjos, J. C., D. Hartill, F. Sauli, and M. Sheaf, *Instrumentation in Elementary Particle Physics*, Rio de Janeiro, World Scientific Publishing Co. (1992).
Fabjan, C. W. and H. F. Fisher, Particle detectors *Repts. Progr. Phys.* **43**, 1003 (1980).
Fabjan, C. W. and J. E. Pilcher, *Instrumentation in Elementary Particle Physics*, Trieste, World Scientific Publishing Co. (1988).
Ferbel, T. *Experimental Techniques in High Energy Physics*, Menlo Park, CA, Addison-Wesley Publishing Co., Inc. (1987).
Green, D. *The Physics of Particle Detectors*, Cambridge, Cambridge University Press (2000).
Jensen, S. *Future Directions in Detector R&D in High Energy Physics in the 1990s – Snowmass 1988*, Singapore, World Scientific Publishing Company (1988).
Kleinknecht, K. *Detectors for Particle Radiation*, Cambridge, Cambridge University Press (1987).
Sauli, F. (ed.) *Instrumentation in High Energy Physics*, Singapore, World Scientific (1992).
Williams, H. H. Design principles of detectors at colliding beams, *Ann. Rev. Nucl. Part. Sci.*, **36**, 361 (1986).

3

Collider physics

It is of the highest importance in the art of detection to be able to recognize out of a number of facts which are incidental and which are vital . . . I would call your attention to the curious incident of the dog in the nighttime. The dog did nothing in the nighttime. That was the curious incident.

Sir Arthur Conan Doyle

Science is the refusal to believe on the basis of hope.

C. P. Snow

In the previous two chapters we first defined the fundamental particles of the SM and their interaction and then discussed how they can be detected and their properties measured. We now know roughly the quality of the measurements we can make. Finally we have given some examples of COMPHEP calculations and this tool is available to us. Thus we are ready to launch the specific study of hadron collider physics.

Now we turn to the question of how particles are produced in proton–(anti)proton (p–p, $\bar{p}$–p) collisions. We will deal only with high transverse momentum, or high mass interactions. There are several reasons for this. The first is that the QCD is weak at high mass scales, and therefore high mass processes can be calculated perturbatively. Secondly, the vast majority of interactions produce particles at low transverse momentum. Thus, the high transverse momentum interactions are the rare ones that stand out above the background. New phenomena can be expected to have a favorable signal to noise ratio in events with particles having a high transverse momentum. Third, if we deal with high mass fundamental interactions, the strong interactions can be "factored out" of the problem, as we will see.

We must define the distribution of quarks and gluons in the initial state proton using experimental data. The dynamics is non-perturbative and therefore is not presently calculable. However, the basic interaction of the SM particles can be predicted for a given process since it is a fundamental process consisting of a point-like interaction between fundamental particles. We will argue that, at high transverse momenta, the basic proton–(anti)proton interaction factorizes into an experimental description of the source of the fundamental particles in the proton, a calculable fundamental process and (perhaps) a second experimental description of the hadronization of the final state fundamental particles into asymptotic, colorless final states.

3.1 Phase space and rapidity – the "plateau"

We begin by looking at the kinematics of the produced, or "secondary" particles. The rapidity variable, y, is defined in Appendix C. The magnitude of the particle momentum is P while energy is E. The momentum component parallel to the beam is labeled by $P_\parallel$, while the perpendicular component is defined to be P_T. The solid angle element is $d\Omega$. The rest mass is m, and the azimuthal angle is ϕ.

$$E = m_T \cosh y$$
$$m_T^2 = m^2 + P_T^2. \tag{3.1}$$

If the transverse momentum is limited by dynamics, we expect (Appendix C) that a particle produced at small y will have a uniform distribution in y if the particle masses are small with respect to the transverse momentum.

As shown in Appendix C, the rapidity, y, is approximated by the pseudorapidity variable, η, defined in Chapter 2. Therefore, the detector shown in Chapter 2 was segmented into "pixels" of equal one particle phase space, $\sim \Delta\eta\Delta\phi$, by design.

As a numerical example, the rapidity of an incident proton in a proton–(anti)proton collision is given below for the Fermilab Tevatron and the CERN LHC. The maximum value of y at fixed E occurs at $P_T = 0$, $\cosh y_{\max} = E/m = \gamma$.

$$pp \text{ at } \sqrt{s} \text{ of } 2, \qquad 14 \text{ TeV}$$
$$\text{has } y_{\max} = 7.7, \qquad 9.6, \tag{3.2}$$

We now give an example of the rapidity "plateau", or region of uniformly distributed y centered on $y = 0$. In this chapter and in later chapters, Monte Carlo results are either the result of "homebuilt" programs written by the author or arise from using the COMPHEP code – running under Windows 2000. More details for COMPHEP are given in Appendix B. These tools are available to the student, empowering a "hands on" exploration by the student using the COMPHEP code.

COMPHEP provides a display of the Feynman diagrams that contribute to the process that is defined by the user, and we will often display them as they help very much in visualizing the nature of the particular problem. A Feynman diagram shows the space-time evolution of the fundamental particles of the SM, which scatter as they exchange the force carriers we discussed in Chapter 1. Space is vertical and time is horizontal in the diagrams given in this text. We show in Fig. 3.1 the fundamental gluon scattering diagrams provided by COMPHEP, where two gluons existing in the incident protons either annihilate to form a single virtual gluon (triplet gluon coupling) or exchange a virtual gluon in analogy to Rutherford scattering.

Proton–(anti)proton scattering has this fundamental process as a sub-process, as we will explain later. For now, we will simply accept the results of the COMPHEP Monte Carlo program for p–p, which are given in Fig. 3.2, and note the existence of a rapidity "plateau" which indicates that the produced particles follow single particle phase space at wide angles.

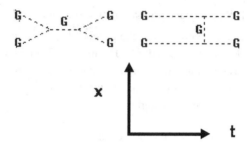

Figure 3.1   The COMPHEP Feynman diagrams for gluon–gluon scattering.

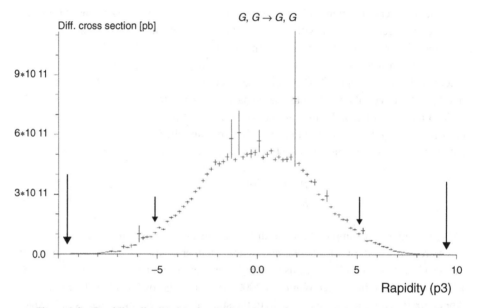

Figure 3.2   Rapidity distribution for produced gluons at the LHC (14 TeV *p–p* CM energy). The small arrows indicate the limits of the angular coverage of the detector shown in Chapter 2. The larger arrows indicate the initial proton beam rapidity in the CM.

Note that the "error bars" shown in the figure are provided by COMPHEP as an estimate of the error in a given data point due to the limited number of Monte Carlo events that are generated. The interested student can run COMPHEP sessions with a variable number of trials, plot the results, and see how the error bars shrink with the longer computations.

The kinematic limit is at rapidity ∼9.6 (final state particle energy cannot exceed the initial particle energy). The region around $y = 0$ (90 degrees in polar angle) has an approximately flat "plateau" with width $\Delta y \sim 6$ for the LHC. Recall the detector coverage out to pseudorapidity of +5 and −5 discussed in Chapter 2. That is, indeed, a good match to the distribution shown in Fig. 3.2. The width of the "plateau" depends on the produced particle mass and transverse momentum (Eq. (3.1)), but only logarithmically. Therefore,

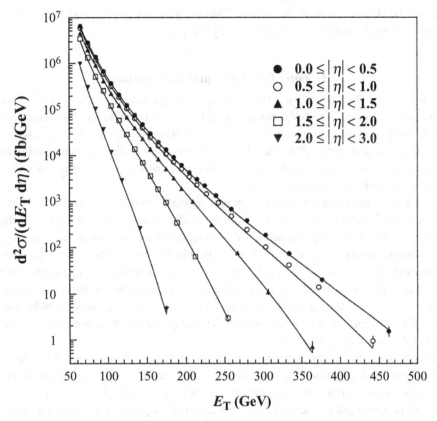

Figure 3.3   D0 data for the jet cross section in different pseudorapidity ranges as a function of transverse energy of the jet ([1] – with permission). The lines represent different distribution function fits.

the plateau width at the LHC will be of order ~6 independent of the dynamics or of the production process, at least for mass scales small with respect to the CM energy.

There are two general purpose experiments in progress at the Fermilab Tevatron accelerator complex, called D0 and CDF. We have already shown examples of events from D0 and CDF in Chapter 2. We will now use data from these experiments to illustrate production. For example, data from the Tevatron experiment D0 are shown in Fig. 3.3. The cross section for the production of "jets" arising from the fragmentation of quarks and gluons is shown as a function of the jet transverse energy for different rapidities. We will use energy and momentum of a jet interchangeably because we assume that jets have negligible masses.

We can easily see that for E_T small with respect to the CM energy of 2 TeV ($E_T \sim$ 100 GeV) there is a rapidity "plateau" at the Tevatron with $dy \sim \pm2$, total width $\Delta y \sim 4$. Comparing LHC (Fig. 3.2) Monte Carlo model predictions and Tevatron data (Fig. 3.3), we see that the plateau width increases with an increase in CM energy. We can also see that the plateau width shrinks at fixed CM energy as transverse energy increases, as expected from the definition of rapidity given in Eq. (3.1).

3.2 Source functions – protons to partons

We will assume that the proton is the incoherent sum (no quantum phases of the wave function) of "valence" u and d quarks, radiated gluons, and a "sea" of quark and antiquark pairs. The proton quantum numbers are satisfied if the proton is thought to be a bound state of $u + u + d$ "valence" quarks. The "sea" gluons can arise from radiation by the valence quarks and the antiquarks can arise from subsequent gluon "splitting" or virtual decay into quark–antiquark pairs.

The lack of interference between quantum amplitudes comes from the fact that there are two fundamental scales to the reaction: the binding energy scale, or the size of the proton, and the "hard" or fundamental collision scale. We will operate at "hard" or large transverse momentum, P_T, scales well above the binding energy scale, $P_T \gg \Lambda_{QCD}$. A proton will disassociate into a virtual state of "partons," or fundamental particles of the SM. This state has a lifetime $\sim 1/\Lambda_{QCD}$, which is long with respect to the collision time that is set by $1/P_T$. During the hard collision, the partons can be considered to be free. Therefore, the partons scatter incoherently and the proton cross section is simply the sum of the individual parton cross sections.

In this limit, the quarks and gluons inside the proton can be represented by classical probability distribution functions. The probability of observing a given constituent of the proton is described by a distribution function, $f_i(x)$ (see Fig. 3.4), where i refers to the type of parton and x is defined to be the fraction of the proton momentum carried by the parton. These distributions are necessarily determined by experiment because they describe the proton binding mechanism at mass scales where QCD is not perturbatively calculable. In this text we will simply accept them as a known input. We assert that the distribution functions are universally applicable to all fundamental processes, as are the fragmentation functions (see Chapter 2) describing the transition from the final state partons to the asymptotic hadron states.

The CM energy of the p–(anti)p state $A + B$ is $\sqrt{s}$. The fundamental "parton" (or point-like particle) reaction is $1 + 2 \rightarrow X \rightarrow 3 + 4$. The fundamental parton dynamics is given a schematic representation as the product of two coupling constants. The first refers to the initial state, $1 + 2$ forces, while the second refers to the final state, $3 + 4$. The two body parton scattering occurs at CM energy (or mass of the composite state X) of $\sqrt{\hat{s}}$. The process is factorized into the distribution of partons in the initial state, the subsequent scattering of those partons, and the final fragmentation of the final state partons into hadrons, if that is applicable.

In what follows we will sequentially examine the different factors from left to right. First we look at the "underlying event" that results from the fragmentation of the fractured proton and (anti)proton after the hard emission of the initial state partons. Then we will consider the distribution functions. In Section 3.3 the initial state $1 + 2$ is explored, followed by the point like scattering, $1 + 2 \rightarrow 3 + 4$ in Section 3.4. The

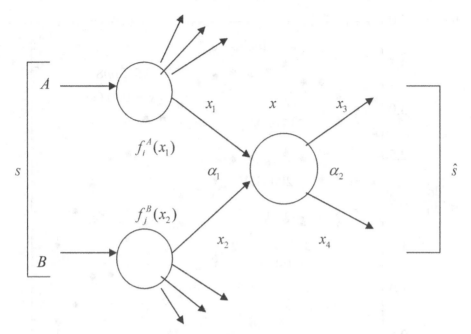

Figure 3.4 Schematic representation of the partons in a proton (A)–(anti)proton (B) collision. The distribution functions for the initial state partons are shown, along with the kinematic definitions of the parton two body scattering and coupling constants.

one and two body final states are then discussed in Sections 3.5 and 3.6 respectively. Fragmentation of the final state partons is considered in Section 3.7, which completes this chapter.

The residual fragments of the fractured p and (anti)p evolve into "soft," $P_T \sim 0.4\,\text{GeV}$, pions with a charged particle density ~ 6 per unit of rapidity and equal numbers of π^+, π^o, and π^-. We have already mentioned the "underlying event" in Chapter 2. We expect that every interaction will contain a similar distribution of "soft," or low transverse momentum particles. In Fig. 3.5 we show the transverse momentum spectrum and the pseudorapidity distribution for the particles produced at low transverse momentum in proton–(anti)proton collisions with no restriction on the final state. The jargon for these events is "minimum bias" events or "inclusive" inelastic interactions, those which occur if no selection, or trigger, on the final state is imposed.

There is clearly a plateau in pseudorapidity with a particle density, which rises slowly with CM energy. The plateau width also increases with CM energy, as expected. The transverse momentum distribution is tightly localized to values $<0.5\,\text{GeV}$. In general, the CM energy dependence for $P_T < 1\,\text{GeV}$ is small. The transverse momentum behavior can be fitted to a power law at low transverse momenta.

$$d\sigma / \pi \, dy \, dp_T^2 \sim A/(p_T + p_0)^n,$$

$$A \sim 450\,\text{mb/GeV}^2, \ p_0 \sim 1.3\,\text{GeV}, \ n \sim 8.2. \tag{3.3}$$

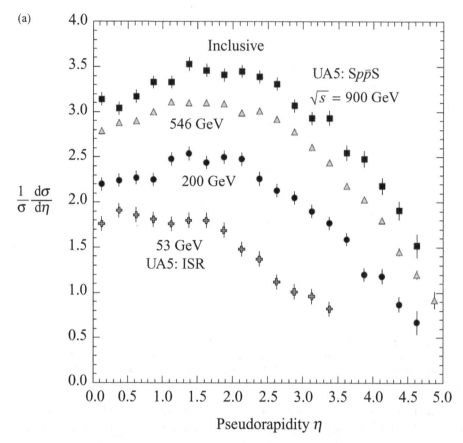

Figure 3.5   (a) Data at different CM energies on the cross section of produced charged
particles as a function of CM pseudorapidity of the particle ([3] – with permission).
(b) Data at different CM energies on the cross section of charged particles produced in
p–(anti)p collisions as a function of their transverse momentum ([2] – with permission).

The coefficient A is of order 100 mb. Since 100 mb is roughly the total inelastic
cross section, the low P_T particles make up the bulk of those produced in an inelastic
interaction in p–p collisions. The falloff of the cross section at transverse momenta well
above ∼p_0 goes as a power of the transverse momentum.

The fragments of hadrons A and B at low P_T merge smoothly with fragmentation
products of "minijets" or jets at "low" P_T for transverse momenta higher than ∼10 GeV.
The production of gluon jets has a cross section of ∼1 mb at a transverse momentum
∼10 GeV. The boundary between the "soft" physics shown in Fig. 3.5 and the "hard
scattering" shown in Fig. 3.6 is not very definite. The Monte Carlo prediction shown
in Fig. 3.6 is a COMPHEP result for gluon–gluon scattering in 14 TeV p–p collisions
(LHC) and does not consider the fragmentation of the gluons.

Leaving the breakup of the fractured p–(anti)p, we now look at the parton distribution functions. We will first try to gain a qualitative understanding of their simplest

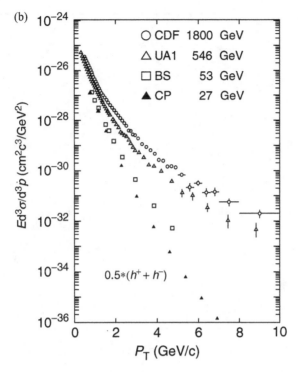

(b)

Ed³σ/d³p (cm²c³/GeV²)

○ CDF 1800 GeV
△ UA1 546 GeV
□ BS   53 GeV
▲ CP   27 GeV

0.5*(h⁺ + h⁻)

P_T (GeV/c)

Figure 3.5   (*cont.*)

characteristics. Suppose first that there was very weak binding of the $u + u + d$ "valence" quarks in the proton. These quarks are the ones that give the proton its quantum numbers, such as charge, $a = e(2/3 + 2/3 - 1/3)$. For weak binding, all three quarks would have the same velocity, as shown in Fig. 3.7.

We expect that the valence quark distribution function, $f(x)$, is a very sharply peaked function centered at $x = 1/3$ in this case. The variable x is the fraction of the momentum of the proton carried by the fundamental particle, or parton. However, the u and d quark masses are ~5 MeV (see Fig. 1.1) and the proton mass is 940 MeV. Therefore the quark motion inside the proton must be relativistic since the effective mass of the total system is much greater than the sum of the masses of the constituents. Since the quarks are bound together in a proton of size ~1 fm, we expect ($\Delta x \Delta P \sim \hbar$, $\Delta x \sim 1$ fm, $P \sim \Delta P \sim 0.2$ GeV $\sim \Lambda_{QCD}$) that they have momenta ~200 MeV, much greater than their rest mass.

Since the bound quarks are in relativistic motion, they can easily radiate gluons. This means that the gluons are distributed, for very small values of x, such that $xg(x) \sim$ constant, where $g(x)$ is the distribution function for gluons. Gluons themselves can then virtually "split" or "decay" into quark–antiquark pairs, which implies that $xs(x) \sim$ constant, where $s(x)$ is the strange quark distribution function. For this reason a distinction is made between the valence quarks and the "sea" of radiated gluons and quark–antiquark pairs (see Fig. 3.9).

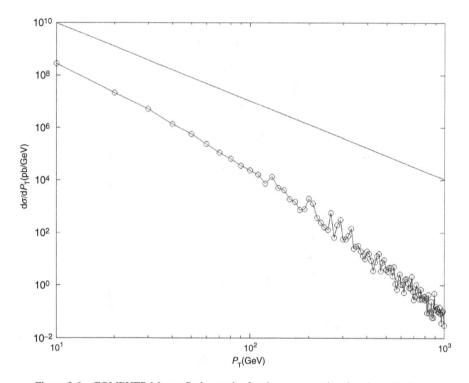

Figure 3.6   COMPHEP Monte Carlo results for the cross section for gluon "jet" production at the LHC at low transverse momentum. The additional solid line indicates
a fundamental cross section, which decreases with transverse momenta as the inverse
cube, $d\sigma/dP_T \sim 1/P_T^3$.

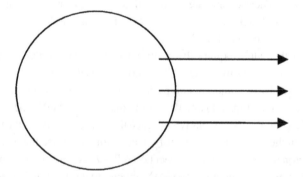

Figure 3.7   Schematic representation of the momentum fraction of the three valence
quarks in a proton. The binding is assumed to be very weak.

We now justify the assertion that $xg(x)$ is constant. The kinematic definitions for the
emission of a massless boson of momentum k, energy ω, by a relativistic fermion of
momentum P are given in Fig. 3.8. The quantity x is defined to be the fraction of the
parent momentum carried off by the boson.

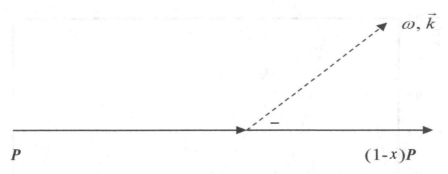

Figure 3.8 Schematic representation of the radiation of a massless particle of energy ω, momentum k, by a particle of momentum P. The final state fermion has a fraction, $1-x$, of the initial fermion momentum.

In perturbation theory the reaction amplitude, A, in non-relativistic quantum mechanics goes as the inverse of the energy difference between the initial and final states (no bosons included) $A \sim 1/\Delta E = 1/(E_f - E_i)$. Therefore, the amplitude for the radiation of a gluon of momentum fraction x goes as $\sim 1/x$, and the emitted gluon will be "soft". We use the approximation that a high energy particle has,

$$E = \sqrt{P^2 + m^2} \approx P + m^2/2P \sim P,$$
$$\Delta E \sim P - (1-x)P,$$
$$A \sim 1/x. \tag{3.4}$$

Using the conservation of both energy, $E = E' + \omega$, and momentum, $\vec{P} = \vec{P}' + \vec{k}$, we assert that, after some considerable algebra, we find the relation $\omega = k \cos\theta = k_\parallel$. Therefore, the massless radiated gluon will be approximately collinear with the parent, $\theta \sim 0$. Radiated gluons are both soft and collinear.

The experimentally determined distribution function of valence quarks, gluons, and sea antiquark–quarks is shown in Fig. 3.9. There is a residual "memory" of the $x \sim 1/3$ value for the valence quarks, but the mean x value is reduced because of radiation. The gluons and sea antiquarks have the characteristic $xf(x) \sim$ constant radiative behavior at small values of x. The gluons are the dominant "partons" at low x values. At larger x values they are highly suppressed and the valence quarks dominate for $x > 0.2$.

Let us briefly mention that the distribution functions depend on the mass scale, Q, at which they are probed. We keep in mind that the variation with mass is slow – logarithmic. To lowest order we could ignore this variation, and we do so for the rest of this chapter. COMPHEP, however, has the appropriate behavior built into the program.

The "running" or variation of basic quantities with mass scale, conventionally called Q, is due to quantum corrections that contain additional powers of the coupling constants. Details are given in Appendix D. The root cause of the "running" behavior in the case of the distribution functions is the radiation by the colored quarks and gluons. For example, a quark with momentum fraction x in the distribution function can be produced by a

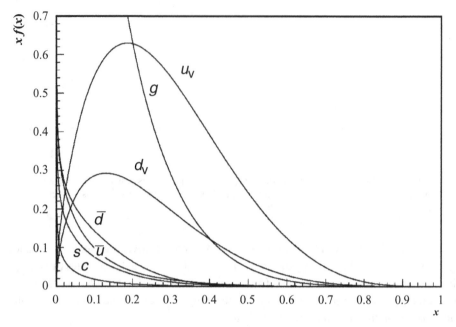

Figure 3.9   Distribution of the momentum fraction of the valence up quarks, gluons, valence down quarks, and "sea" antiquarks in a proton ([4] – with permission). The curves are fits at a fixed momentum transfer scale (see Appendix D).

quark at a higher momentum fraction that has subsequently radiated a gluon and thus lost energy (see Fig. 3.8).

QCD perturbation theory provides us with the description of the emission of a quark plus gluon by a quark. In principle we could now "evolve" the distribution functions, $f(x,Q^2)$, from one mass scale, Q, to any other mass scale by solving a set of equations describing all the radiative processes that quarks and gluons undergo. The result is that as the mass scale increases the importance of radiative processes grows, which enhances all the distribution functions at lower x, depleting them at high x. For example, the gluon distribution grows rapidly at low x as Q increases. This behavior is seen in Fig. 3.10 for $x < 0.02$, where $g(x)$ grows faster than $1/x$.

COMPHEP has two sets (MRS and CTEQ) of available distribution functions. It is advisable for the student to run the program for the same process but using the two different distribution functions. If they are well measured in the region of x and Q^2 probed by the process in question, the results should be insensitive to the choice of distribution function. If they are not, then there is a "theoretical" uncertainty in the predicted cross section because the distribution functions have been extrapolated to regions of the parameter space beyond where they have been well measured.

Gluons are observed in lepton scattering experiments to carry approximately half the proton momentum. That fact can be used to normalize the gluon distribution. A suppression at high x values is accomplished by assuming a $(1 - x)^6$ factor in the $xg(x)$

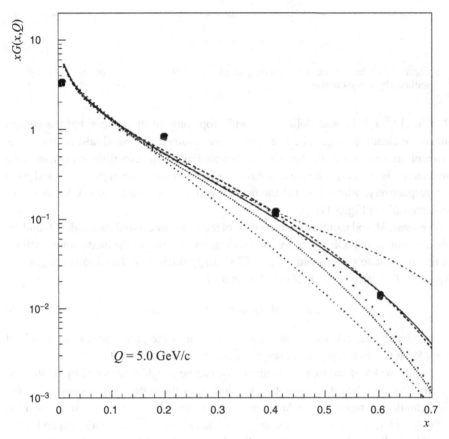

Figure 3.10   Gluon distribution functions taken from different fits to experimental data, shown as lines. The circles are a few points from Eq. (3.5) ([5] – with permission).

distribution:

$$xg(x) = 7/2(1 - x)^6,$$

$$\int xg(x)\mathrm{d}x = 1/2. \tag{3.5}$$

Some fits representing the measured gluon distribution function are shown in Fig. 3.10. The discrete points are representative values of Eq. (3.5), showing that this simple parameterization is a reasonable first approximation to the gluon distribution. Therefore, for gluon induced reactions we can also have confidence in our ability to make a "back of the envelope" calculation.

3.3 Two body formation kinematics

The parton distribution functions give us the joint probability of finding a parton of type i at momentum fraction x_1 emitted by hadron A and parton of type j at x_2 from hadron

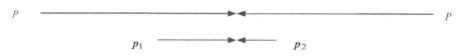

Figure 3.11   Schematic representation of the initial state in parton–parton scattering initiated by a p–p collision.

B, $f_i^A(x_1)f_j^B(x_2)$. In what follows we will drop some of the indices, but the context should be clear (see Fig. 3.4). The partons are assumed to have almost no transverse momentum, since we argue that the scale for binding energy contributions to transverse momentum is $\sim \Lambda_{\text{QCD}}$. The partons have longitudinal momentum $p_1 = x_1 P$ and $p_2 = x_2 P$ respectively, where P is the momentum of the proton in the p–p CM, as shown schematically in Fig. 3.11.

The mass, M, and momentum fraction, x, of the composite initial state is then found, by conservation of relativistic energy and momentum, in terms of the momentum fractions of the initial state partons and the p–p CM energy squared, s. The details are given in Appendix C, but $x = x_1 - x_2$ should be obvious.

$$x_1 x_2 = M^2/s \equiv \tau, \quad x = x_1 - x_2. \tag{3.6}$$

A typical value, $\langle x \rangle$, for the momentum fraction of the parton producing a state of mass M (at $x = 0$) at a p–p CM energy $\sqrt{s}$ is $\sqrt{\tau} = M/\sqrt{s}$.

The full width of the rapidity plateau, Δy, can be roughly estimated by finding the kinematic limit when the momentum fraction, x, of the system approaches 1. We use the definition of rapidity (see Appendix C), $E = m_T \cosh y$, $P_{\parallel} = m_T \sinh y$ and the definition of x, $x = P_{\parallel}/P = m_T \sinh y/P = 2m_T \sinh y/\sqrt{s}$. The width depends only logarithmically on the mass of the produced state and the CM energy. Note that $x = 1$ implies $y = y_{\text{max}}$ and $\Delta y = 2y_{\text{max}}$.

$$x = (2m_T \sinh y/\sqrt{s}) \sim (M/\sqrt{s})e^y$$
$$\Delta y \sim 2\ln(\sqrt{s}/M). \tag{3.7}$$

A system of mass M is formed by a parton with x_1 from proton A and a parton with x_2 from (anti)proton B. The joint probability, $P_A P_B$, to form a system of mass M moving with momentum fraction x assumes independent emission of the two partons. The variable C in Eq. (3.8) is a color factor having to do with normalization of the distribution functions, which we will explain later, as needed. The fundamental parton scattering is described by the cross section $\hat{\sigma}$ while the proton–(anti)proton cross section is σ.

$$d\sigma = P_A P_B d\hat{\sigma} = C f^A(x_1) dx_1 f^B(x_2) dx_2 d\hat{\sigma} (1 + 2 \to 3 + 4). \tag{3.8}$$

We make a change of variables in order to express the cross section in terms of observables in the final state, M and y, converting from x_1 and x_2, $dx_1 \, dx_2 = d\tau \, dy$. Once we measure M and y in the detector, we can infer the values of x_1 and x_2, at least for two body scattering (see Appendix C).

$$d\sigma = Cf^A(x_1)f^B(x_2)d\tau \, dy \, d\hat{\sigma}(1 + 2 \to 3 + 4),$$
$$(d\sigma/d\tau \, dy)_{y=0} = Cf^A(\sqrt{\tau})f^B(\sqrt{\tau})d\hat{\sigma}(1 + 2 \to 3 + 4). \tag{3.9}$$

Assuming a plateau of width Δy we can estimate the full cross section as follows: $\Delta\sigma \sim (d\sigma/dy)_{y=0}\Delta y$. The value of Δy varies only slowly with mass (see Eq. (3.7)), and is a number of order 4–10 at the LHC.

The last line of Eq. (3.9) shows that the differential cross section is a function of a single dimensionless variable, τ. This is an immediate prediction of the model, independent of any particular dynamical assumptions. This "scaling" behavior is confirmed in a wide variety of hadron collider data. An example using jet data at two different CM energies is shown later in this chapter. We see also that in order to make further progress we must know the fundamental scattering process, $d\hat{\sigma}(1 + 2 \to 3 + 4)$. We know how to compute this scattering cross section since we believe we understand the dynamics of the fundamental particles of the Standard Model.

3.4 Point like scattering of partons

We are now, moving left to right in Fig. 3.4, at the point of considering the fundamental parton scattering process. In non-relativistic quantum mechanics, the Born approximation to the amplitude, A, for a process is the interaction Hamiltonian sandwiched between initial and final plane wave (free particle) states $|i\rangle$ and $|f\rangle$, $A = \langle f|H_1|i\rangle \sim \int e^{i\vec{q}\cdot\vec{r}}V_1(r)d\vec{r}$, which is just the Fourier transform of the interaction potential, $V_1(r)$, where $\vec{q} = \vec{k}_f - \vec{k}_i$, $q \sim k\theta$ is the magnitude of the momentum transfer in the reaction. A familiar example is the $1/r$ Coulomb potential, which yields a Born amplitude $\sim 1/q^2$ describing how the virtual exchanged photon propagates in momentum space. In turn this leads to a cross section (Rutherford scattering) that goes as the square of the amplitude $\sim 1/q^4 \sim 1/\theta^4$, which should be familiar.

We use the relativistic parton variable $\hat{s}$, the CM energy squared, and $\hat{t}$, the four-dimensional momentum transfer, $(p_3 - p_1)_\mu \cdot (p_3 - p_1)^\mu$. The variable $\hat{u}$ is defined such that $\hat{s} + \hat{t} + \hat{u} = 0$, ignoring the small masses of the partons. The point like cross section we use has an overall factor that contains the coupling constants at the two vertices in Fig. 3.4 called out explicitly as well as the general point like energy dependence.

$$\hat{\sigma} \sim \pi(\alpha_1\alpha_2)|A|^2/\hat{s}. \tag{3.10}$$

The remaining factors are dimensionless and depend on the specific process as given in Table 3.1, where the $\hat{s}$ and $\hat{t}$, "hat" notation is dropped for simplicity. The entries are all numbers of order unity at large scattering angles, $\hat{\theta} = \pi/2$. The $1/t^2$ behavior, $t \sim q^2$, expected in Rutherford scattering is also in evidence. Therefore, the expression for a general point like cross section given in Eq. (3.10) is a useful first approximation to the cross section. We will adopt it in making our "back of the envelope" calculations.

Table 3.1 *Point like cross sections for parton–parton scattering. The entries have the generic dependence of Eq. (3.10) already factored out. At large transverse momenta, or scattering angles near 90 degrees (y ∼ 0), the remaining factors are dimensionless numbers of order one ([4] – with permission).*

Process	$|A|^2$	Value at $\theta = \pi/2$
$q + q' \rightarrow q + q'$	$\frac{4}{9}[s^2 + u^2]/t^2$	2.22
$q + q \rightarrow q + q$	$\frac{4}{9}[(s^2 + u^2)/t^2 + (s^2 + t^2)/u^2] - \frac{8}{27}(s^2/ut)$	3.26
$q + \bar{q} \rightarrow q' + \bar{q}'$	$\frac{4}{9}[t^2 + u^2]/s^2$	0.22
$q + \bar{q} \rightarrow q + \bar{q}$	$\frac{4}{9}[(s^2 + u^2)/t^2 + (t^2 + u^2)/s^2] - \frac{8}{27}(u^2/st)$	2.59
$q + \bar{q} \rightarrow g + g$	$\frac{32}{27}[t^2 + u^2]/tu - \frac{8}{3}[t^2 + u^2]/s^2$	1.04
$g + g \rightarrow q + \bar{q}$	$\frac{1}{6}[t^2 + u^2]/tu - \frac{3}{8}[t^2 + u^2]/s^2$	0.15
$g + q \rightarrow g + q$	$-\frac{4}{9}[s^2 + u^2]/su + [u^2 + s^2]/t^2$	6.11
$g + g \rightarrow g + g$	$\frac{9}{2}[3 - tu/s^2 - su/t^2 - st/u^2]$	30.4
$q + \bar{q} \rightarrow \gamma + g$	$\frac{8}{9}[t^2 + u^2]/tu$	
$g + q \rightarrow \gamma + q$	$-\frac{1}{3}[s^2 + u^2]/su$	

These estimates should be made as a "reality check" before jumping into the COMPHEP program.

We define the luminosity, L, such that the luminosity times the cross section, σ, gives the observed interaction rate in reactions per second. As an example, the LHC has a design luminosity leading to a total inelastic interaction rate of ~ 1 GHz. Since the accelerator has radio frequency (rf) bunched beams crossing every 25 ns, there are ~ 25 inelastic interactions contained in each bunch crossing. This leads to "pileup" in a detector since events within a bunch crossing cannot be temporally resolved.

$$\sigma \sim 100 \text{ mb} = 10^{-25} \text{ cm}^2,$$
$$L \sim 10^{34}/(\text{cm}^2 \text{ s}),$$
$$\sigma L \sim 10^9 \text{ Hz.} \tag{3.11}$$

As a quick "reality check," we revisit the low transverse momentum jet rates. Because the process occurs at low mass and hence small x, the gluon–gluon cross section dominates. The probability of finding a small P_T jet, or "minijet," in an LHC crossing is not small. We estimate in Eq. (3.12) the cross section for producing gluon pairs above a mass

M_o from Eq. (3.9) and Eq. (3.10).

$$M^3(d\sigma/dM\,dy)_{y=0} = 2[xg(x)]^2 C(d\hat\sigma\hat{s})(\hbar c)^2,$$
$$\Delta\sigma(M > M_o) \sim \Delta y[xg(x)]^2[\pi\alpha_s^2|A|^2/M_o^2](\hbar c)^2. \tag{3.12}$$

The differential cross section falls with mass as the third power. This power law behavior is characteristic of point like fundamental processes. We can use the gluon distribution normalization, the rapidity full width, and the strong coupling constants to estimate the jet–jet cross section for masses >10 GeV. For small x, $[xg(x)] \sim 7/2$. The rapidity width is ~ 10, while $\alpha_s \sim 0.1$. Using $|A(g+g\to g+g)|^2$ from Table 3.1, we find a cross section ~ 0.4 mb above a mass of 10 GeV.

It is a rough "reality check" of Fig. 3.6 with $M/2 \sim P_T$ (see Appendix C), that the simple estimate of the cross section is a number of order 1 mb. We took $C \sim 1$ which means we ignored the color matching of the gluon from hadron A to that from hadron B. We are then assuming that any color mismatch can be radiated away by very soft gluons with probability about one which does not reduce the reaction rate.

3.5 Two-to-one Drell–Yan processes

We are now going to look at resonant formation of a single particle in the final state. For historical reasons this is called "Drell–Yan" production. We first recall that in quantum mechanics a resonance describes an unstable state with a mass, M, and a distribution, the Breit–Wigner distribution, of masses having a finite width, Γ. The decaying state then has a finite lifetime $\tau \sim \hbar/\Gamma$. The cross section for producing a state of spin J is limited by unitarity, $\hat\sigma < 4\pi\lambda_{dB}^2(2J+1)$, where the deBroglie wavelength, λ_{dB}, is related to the CM momentum, $\hat{P}$, and hence the mass, M, $\lambda_{dB} \sim \hbar/\hat{P} \sim 2\hbar/M$.

We will assume that the width is small with respect to the mass, and then integrate the CM energy over a mass range roughly equal to the width of the resonance. In this way, we integrate Eq. (3.9) over the final state mass to find the cross section for resonance production as a function of rapidity. The partial width for formation of the state in the reaction $1+2$ is defined to be Γ_{12}.

$$\int(\hat\sigma)d\hat{s} = \pi^2(2J+1)(\Gamma_{12}/M),$$
$$M^2(d\sigma/dy)_{y=0} = C[xf^A(x)xf^B(x)]_{x=\sqrt\tau}[\pi^2\Gamma_{12}(2J+1)/M]. \tag{3.13}$$

In order to obtain a rough estimate of the cross section we note in the absence of any dynamics the ratio of the resonant width to the mass is defined by the strength of the relevant coupling constant, α_{12}:

$$\Gamma_{12}/M \sim \text{``}\alpha_{12}\text{''}. \tag{3.14}$$

The cross section on the plateau times the square of the mass also "scales." It is a function only of the dimensionless variable τ. This predicted behavior has been observed

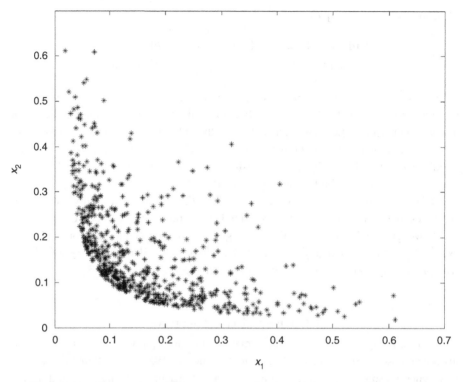

Figure 3.12 Scatter plot of the momentum fraction of the gluons in a proton–(anti)proton collision. The produced mass is fixed at 200 GeV and the overall CM energy is 2 TeV.

in, for example, the production of W and Z bosons at different CM energies. As a rough estimate we expect the cross section to be $\sigma_{12} \sim \Gamma_{12}/M^3 \sim \alpha_{12}/M^2$.

Let us look at the kinematic correlation between the two partons in the initial state. A simple Monte Carlo program has been written that picks x_1 and x_2 from $xg(x)$, weighting by the dynamics, $\sim 1/M^2$ (see Eq. (3.13)). The final state mass is fixed at 200 GeV and the CM energy is 2 TeV. The scatter plot of the accepted x values is shown in Fig. 3.12. There is a hyperbolic kinematic boundary, where $\langle x \rangle \sim 0.1$, which is the $y = 0$ value occurring when $x_1 = x_2$. Because we produce a fixed mass the kinematic boundary, $x_1 x_2 = M^2/s = 0.01$, is quite sharp. The minimum value of the momentum fraction of one parton occurs when the other parton has an x value of 1, $x_{min} = \tau = M^2/s$. In this case the minimum value is $x = 0.01$.

Now let us look at the production of single W and Z gauge bosons as a function of the available CM energy. The W and Z couple to the u and d quarks in the proton, since the gluon has no flavor or weak charge. Therefore, the production mechanism arises from quarks and antiquarks in the initial state. There is no sharp proton "threshold" energy for W production because the quarks have a wide distribution of momenta within the proton. We can think of the proton as a beam of quarks and gluons with a broad momentum range.

Figure 3.13 Feynman diagrams given in COMPHEP for the production of W and Z gauge bosons.

The COMPHEP Feynman diagrams for these production processes are shown in Fig. 3.13. The W and Z are formed in the reactions $\bar{u} + u \rightarrow Z \rightarrow e^+ + e^-$, $\bar{u} + d \rightarrow W^- \rightarrow e^- + \bar{\nu}_e$. Incidentally, COMPHEP does not allow single particles in the final state, which is why we chose a particular W and Z decay mode.

In COMPHEP upper case indicates an antiparticle (see Appendix B). The initial state contains a quark–antiquark pair, while the final state has a lepton and an antilepton. The coupling of quarks and leptons to gauge bosons is familiar from the discussion in Chapter 1 and Appendix A.

We will use here, and later, the up quarks alone as a rough first estimate of the cross section, because electromagnetic cross sections go as the square of the quark charges. Thus the up dominates over down quarks in the cross section sum by a factor of four. The student should try different quark–antiquark pairs in the initial state in COMPHEP for Z production to verify this assertion. In principle we should use COMPHEP for each possible initial state and add the results incoherently.

At a fixed resonant mass, M, we expect that there is a rapid rise of the cross section with increase in the CM energy, due to the rapid increase in the quark distribution functions with a decrease in the average x value of the distribution functions which is sampled in the reaction, $\langle x \rangle \sim M/\sqrt{s}$. The COMPHEP results are shown in Fig. 3.14. The cross section is substantial, $\sigma_W \sim 30$ nb (we used $B(W^- \rightarrow e^- + \bar{\nu}_e) \sim 1/9$ – see Chapter 4) at the LHC. The "absolute" threshold, when both partons have $x \sim 1$, $\sqrt{s} = M_W = 80$ GeV, is very suppressed because the source distributions vanish there.

The cross section rises by a factor of ten going from the Fermilab Tevatron to the LHC. Even at the LHC the W cross section is only one part in three million of the total inelastic cross section. Clearly there is a premium on efficient and incisive triggering of the detector prior to storage of a candidate event to permanent media.

In Appendix A, we show how the coupling of the Z boson to fermions depends on the Weinberg angle. We also comment that this angle was experimentally determined from data taken in "neutral current" or Z mediated neutrino interactions. The possibility also exists of determining this angle from examining Drell–Yan production of lepton pairs at proton–antiproton colliders such as the Tevatron. In this way, the W mass, the top mass, and the Weinberg angle can all be measured in a single experiment, thus reducing possible systematic effects that might arise in combining data taken by different experiments at different accelerators.

The forward–backward angular asymmetry in quark–antiquark annihilations to electron–positron pairs is shown in Fig. 3.15. The student can easily check these results using COMPHEP.

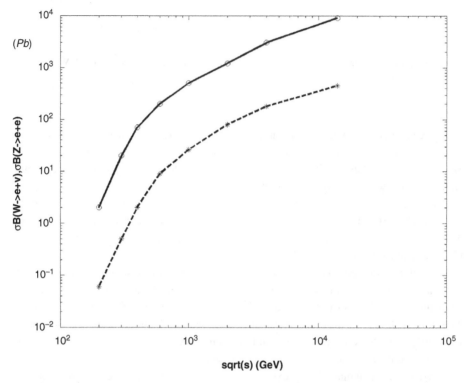

Figure 3.14   COMPHEP results for the cross section times electron branching ratio in the production of W (o, solid) and Z (*, dashed) gauge bosons as a function of the proton–proton CM energy for the fundamental processes shown in Fig. 3.13.

To date, the Tevatron luminosity has been insufficient to acquire enough Z leptonic decay events to make a precise measurement of the Weinberg angle. In future Tevatron data taking, the expected Z statistics will be sufficient. Present data from CDF on the asymmetry are shown in Fig. 3.16. The large value of the asymmetry near the Z mass is due to the different V-A coupling of the L and R quark components to the Z, as discussed in Appendix A. The possible existence of new higher mass Z bosons not present in the SM might be seen in the appearance of a similar structure in the asymmetry at high mass.

There are other processes leading to the production of a single resonant state. The charmed quarks introduced in Chapter 1 can form charm–anticharm bound states before the charmed quarks decay. These states are called charmonium, in analogy to positronium, and an example was shown in Chapter 2 where a charmonium resonance was used in calibrating muon detectors. These resonances have extremely narrow natural widths because they decay by multiple gluon emission, rather like the slow multi-photon decays of ortho and para positronium, which is the electron–positron bound state.

The charmonium states are readily formed in p–p collisions using the gluons contained in the protons. These states are usually detected using their two-lepton decay modes since leptons are rare and thus are readily triggered on. Data are shown in Fig. 3.17 on the transverse momentum of the produced charmonium states.

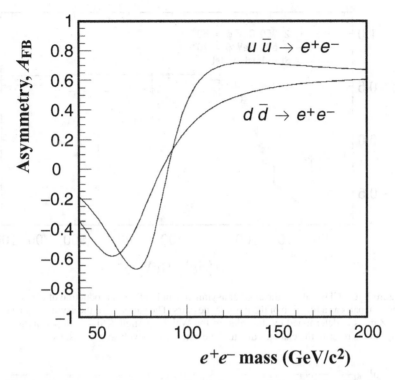

Figure 3.15 Decay angular asymmetry in quark–antiquark annihilations. Interference effects arise because there are two amplitudes with different phases, one with an intermediate photon and one with a Z boson ([5] – with permission). The u and d quarks couple differently to the Z boson (see Appendix A).

The scale set by the transverse momentum distribution of the charmonium system is only a few GeV. As we already argued, the initial state has a very limited transverse momentum set by the characteristic QCD energy scale. The data shown above serve to validate the assumption that the transverse momentum of the initial state is small.

At higher order in the coupling constants this simple picture becomes more complex. The process called "initial state radiation," see Fig. 3.18, where a gluon radiates a gluon prior to the charmonium, Ψ, formation also gives transverse momentum to the charmonium in the final state. Finite values of P_T arise from both initial state radiation and the intrinsic parton transverse momentum.

Let us try to roughly estimate the cross section that we observe in Figs. 3.14 and 3.17. The parton level cross section has previously been quoted in Eq. (3.16). We estimate a W cross section $\hat{\sigma} \sim \pi^2 \Gamma (2J + 1)/M^3$, using a "generic" width ~ 2 GeV ($\sim \alpha_W M$), and obtain $\hat{\sigma} = 47$ nb. This is in reasonable agreement with the full COMPHEP calculation, Fig. 3.14. For charmonium, whose width is only $0.000\,087$ GeV, with a mass of 3.1 GeV, we similarly estimate the cross section $\hat{\sigma}$ to be 34 nb, which is also in rough agreement with the data, Fig. 3.17. The formulae given in this chapter for Drell–Yan production are

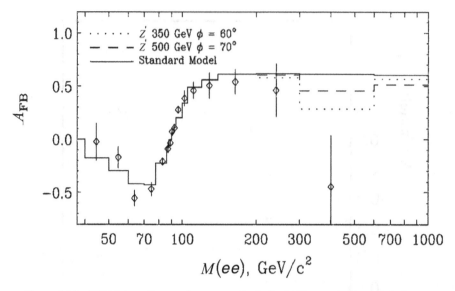

Figure 3.16   CDF data on the angular asymmetry in Drell–Yan production of electron–positron pairs as a function of the mass of the pairs. The variation of the asymmetry near the Z mass is determined by the value of the Weinberg angle ([6] – with permission). The curves refer to the changes due to the existence of new heavy Z′ bosons.

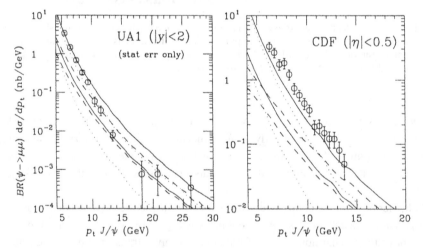

Figure 3.17   Transverse momentum distribution for the production of charmonium states at the UA1 (CERN) and CDF (Tevatron) experiments ([7] – with permission).

therefore validated as a useful first approximation. Note that COMPHEP is incapable of handling charmonium because it only calculates fundamental processes.

We can expand the discussion to look at the production of pairs of particles. In Fig. 3.19 we show the cross section for the production of Z boson pairs as a function of the CM energy. The COMPHEP results show a steep rise with CM energy.

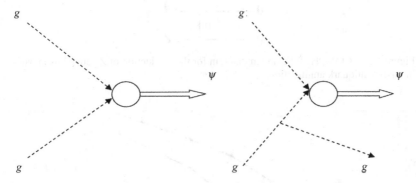

Figure 3.18 Schematic representation of the gluon–gluon formation of charmonium. The emission of an additional gluon leads to a small transverse momentum for the recoiling charmonium.

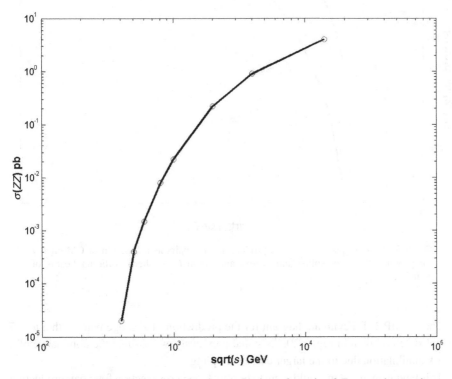

Figure 3.19 COMPHEP results for the production of a pair of Z gauge bosons in proton–proton collisions as a function of CM energy for u quark annihilation in the initial state.

There is a twenty-fold rise in the cross section from the Tevatron to the LHC. Nevertheless, the cross section for ZZ is still only ~ 8 pb at the LHC. Therefore, a high luminosity is necessary even at the high CM energy available at the LHC if we wish to study gauge boson pairs with high statistics.

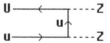

Figure 3.20   COMPHEP Feynman diagram for the production of Z gauge boson pairs in quark–antiquark annihilation.

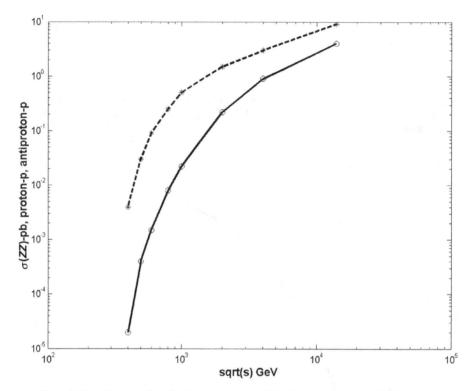

Figure 3.21   Cross section for the production of Z pairs as a function of CM energy for proton–proton (o, solid) and proton–antiproton (*, dashed) colliding beams of hadrons.

The COMPHEP Feynman diagram for the production of Z gauge pairs with a $u + \bar{u}$ initial state is shown in Fig. 3.20. As stated previously, we assume the dominance of u quark annihilation due to the larger charge coupling.

This Feynman diagram would seem to imply a larger cross section for Z pair production in proton–antiproton interactions rather than in p–p interactions since in the former case there are valence antiquarks available. However, this is only true if the typical x value of the distribution functions is large, favoring valence partons. For example, at a CM hadron–hadron energy of 0.4 TeV, the average x is $\langle x \rangle \sim 2M_Z/\sqrt{s}$ or ~ 0.46, where the partons are dominated by valence sources and thus we expect the p–$\bar{p}$ cross section to exceed the p–p cross section. A COMPHEP comparison of Z pair production in

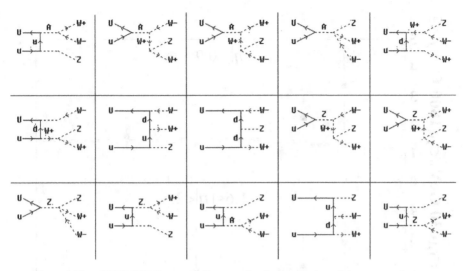

Figure 3.22 COMPHEP Feynman diagrams for the production of three gauge bosons, $W + W + Z$.

proton–proton and proton–antiproton interactions as a function of CM energy is shown in Fig. 3.21.

There is a large difference in cross section at low CM energies (high x values where valence partons dominate), which decreases as the CM energy increases. At LHC energies we expect a factor of less than two difference, which is more than compensated for by the ability to produce high luminosity beams in the p–p case. Basically, if we are in the "sea," a proton is as good as an antiproton for the production of new particles.

Gauge pairs will be discussed further in Chapter 5 in the context of the search for the Higgs boson. The gauge bosons are predicted to have both triplet and quartic self-couplings (see Appendix A). Therefore, we also expect the production of three gauge bosons. The COMPHEP Feynman diagrams appropriate to the production of three gauge bosons, $W + W + Z$, in u quark annihilations are shown in Fig. 3.22.

The diagrams contain vertices with both triple and quartic couplings. Clearly, it is important to explore the production of both gauge boson pairs and three gauge bosons in order to understand whether the triplet and quartic couplings are measured to be what the Standard Model predicts. This study will be an active part of the LHC research program. At present, the achieved luminosities at the Tevatron have not been sufficient to study gauge boson pairs in any detail.

3.6 Two-to-two decay kinematics – "back to back"

We now turn explicitly to the production of two particles in the final state. This is the general case of "two-to-two" scattering. The generic results are shown in Eq. (3.15). On the rapidity plateau, $y \sim 0$, we again expect a scaling distribution; the cross section for

(a)

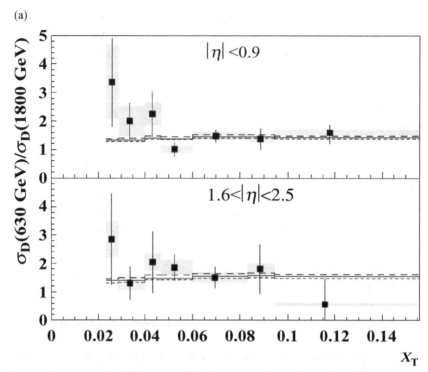

Figure 3.23 $P_T^3 d\sigma/dP_T dy$, the scaling cross section, is compared as a function of the $x_T = 2P_T/\sqrt{s} \sim M/\sqrt{s}$ variable at low x_T for (a) inclusive photons ([9] – with permission) and (b) inclusive jets ([8] – with permission) at the Tevatron in the D0 experiment. The lines are fits to different Monte Carlo models, while the shading indicates systematic errors in the data.

two body scattering depends on a single variable, τ.

$$M^3(d\sigma/dy\,dM)_{y=0} = 2C[xf^A(x)xf^B(x)]_{x=\sqrt{\tau}}(d\hat\sigma\hat s),$$

$$d\hat\sigma \approx \pi\alpha_1\alpha_2/\hat s,$$

$$M^4(d\sigma/dy\,dM^2)_{y=0} \sim C[xf^A(x)xf^B(x)]_{x=\sqrt{\tau}}(\pi\alpha_1\alpha_2). \qquad (3.15)$$

In Fig. 3.23, data taken by D0 on the production of inclusive jets and prompt photons at two different energies are compared with the scaling expectation. The single jet variable used is $x_T = 2P_T/\sqrt{s} \sim M/\sqrt{s}$, which is approximately the scaling variable, $\sqrt{\tau}$. Indeed, the data are roughly only a function of that single scaling variable, thus confirming the prediction. However, exact scaling cannot be true due to the evolution of the source distribution functions with changes in mass scale $Q \sim M$. Therefore, the Tevatron data on jets and photons serve, in their fine details, as a confirmation of the expectation of scaling behavior, modified by corrections due to evolution, which amount to factors of about 1.5–1.7.

(b)

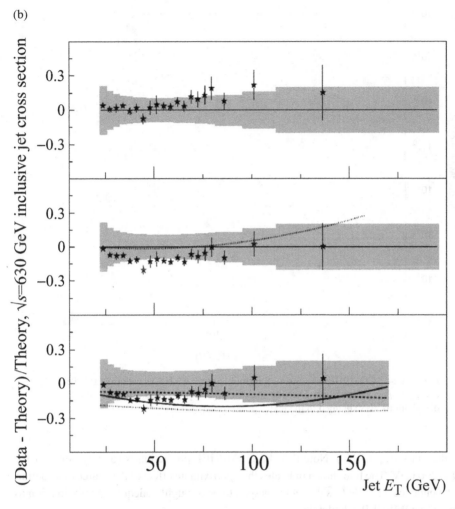

Figure 3.23   *(cont.)*

We expect $1/M^3$ behavior of the cross section as a function of mass, $d\sigma/dM$, at low mass where the parton distribution functions have a slow variation with x. This behavior is a reflection of the power law two body behavior of the generic parton–parton scattering cross section. When $M/\sqrt{s}$ becomes substantial, the source effects will become large. As a numerical example, for $M = 400$ GeV, at the Tevatron, $M/\sqrt{s} = 0.2$, and the factor $(1-M/\sqrt{s})^{12}$, approximating the product of the two gluon distributions, is ~ 0.07. We want to see if we can estimate the falloff of $M^3 d\sigma/dM$ accurately because this quantity reflects the distribution functions.

In Fig. 3.24 we show COMPHEP Monte Carlo model predictions for the distributions of jet–jet mass at a center of mass energy of 2 TeV. We have already removed the expected behavior of the parton–parton cross section by multiplying the cross section by the cube

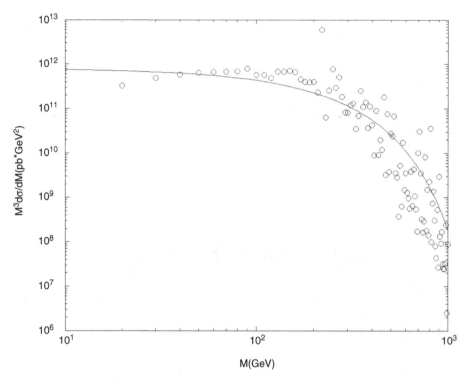

Figure 3.24   COMPHEP results, o, for gluon–gluon two body scattering at 2 TeV CM energy in *p*–*p* interactions. The line indicates the approximate effect of gluon source distributions, as explained in the text.

of the mass (Eq. (3.16)). Note that the COMPHEP prediction is roughly constant for $M < 200$ GeV. The line shown indicates the approximate effect of the source distribution *x* dependence, $(1 - M/\sqrt{s})^{12}$, which is seen to be a roughly adequate approximation to the full COMPHEP calculation.

In Fig. 3.25 we show Monte Carlo COMPHEP predictions for the distribution of jet transverse energy and jet–jet mass at 2 TeV CM energy, with jet rapidity less than 2. As mentioned above, we have the approximate kinematic relationship, $P_T = (M/2)\sin\hat{\theta} \sim M/2$, for large scattering angles. Thus the value of the cross section at a given mass is approximately the value at a transverse energy one half that value, as indicated by the scale chosen in the two figures. As before, we observe an approximate $[1/M^3][1-M/\sqrt{s}]^{12}$ behavior of the mass distribution.

We can look at interactions other than simple strong production. For example, we can look at the production of photons in proton–(anti)proton collisions. The basic COMPHEP Feynman diagrams for this process are shown in Fig. 3.26. We must now have a quark in the initial state because photons couple to the charge of the quark, while the gluons have no electric charge. We also want a gluon in the initial state, as it is the most probable parton at low *x*.

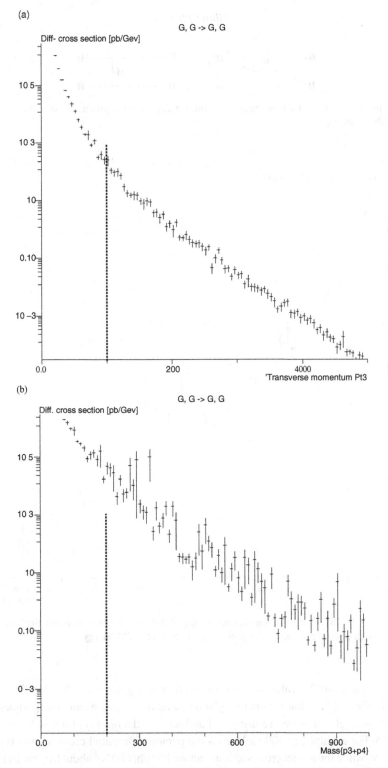

Figure 3.25 COMPHEP results for two body gluon–gluon scattering at 2 TeV CM energy in p–p collisions. (a) gluon (jet) transverse momentum distribution (b) gluon–gluon (jet–jet = dijet) mass distribution. The dotted lines indicate the same cross section level at a mass of 200 GeV and a transverse momentum of 100 GeV.

Figure 3.26   COMPHEP Feynman diagrams for single photon production due to quark–gluon scattering.

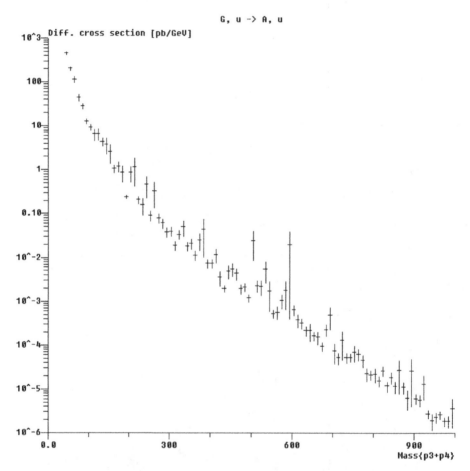

Figure 3.27   Mass distribution obtained by COMPHEP for the photon–quark final state in prompt photon production for *p–p* collisions at 2 TeV CM energy.

We can compare the value of the cross section at a given mass for this final state, shown in Fig. 3.27, to that for the two gluon (we assume a gluon can be experimentally observed as a jet and therefore use gluon and jet interchangeably) final state shown in Fig. 3.25. For example, at 300 GeV mass the photon differential cross section is about 2 pb/GeV, while the jet–jet cross section is about 100 pb/GeV or about fifty times larger. We expect a similar shape for the mass distribution because all the point like differential

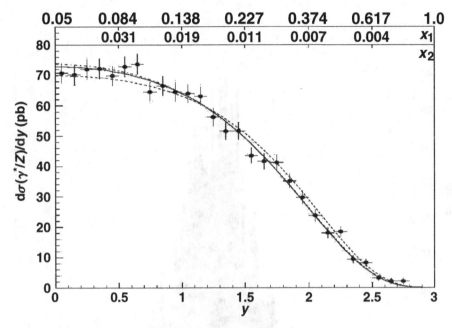

Figure 3.28 CDF data on the Drell–Yan production of electron–positron pairs with a mass approximating the Z mass. The lines are fits to different quark distribution functions. The two body final state variables are used to find the momentum fractions of the two partons in the initial state ([10] – with permission).

cross sections have similar behavior, see Table 3.1. This similarity is at least qualitatively observed.

The rate for prompt photons is expected to be reduced with respect to the jet–jet rate by the ratio of the coupling constants (electromagnetic to strong) and by the differences in the u and g source functions. Those two factors are roughly $\alpha/\alpha_s \sim \frac{1}{14}$ and $u/g \sim \frac{1}{6}$ at $x \sim 0$ (see Fig. 3.9) leading to a net factor of 64. Thus we can crudely understand the ratio of the cross sections.

We now turn to the scattering and the detection of the two body final state. The kinematic details are explained in Appendix C. Suffice it to say that using the measured values of the two final state jet kinematic quantities, rapidity, y_3, y_4, and E_T allows us to solve for x, M, and the CM scattering angle $\hat{\theta}$. Further, we can relate M, y_3, and y_4 to the initial state momentum fractions x_1 and x_2, thus completely specifying the kinematics for the two body process:

$$x_1 = [M/\sqrt{s}]e^y, \; y = (y_3 + y_4)/2,$$
$$x_2 = [M/\sqrt{s}]e^{-y}. \tag{3.16}$$

Data from CDF on the Drell–Yan production of lepton pairs at 2 TeV CM energy are shown in Fig. 3.28. The values of the initial state parton x values are also given in the figure. Note also the nice illustration of the rapidity plateau in this process.

(a)

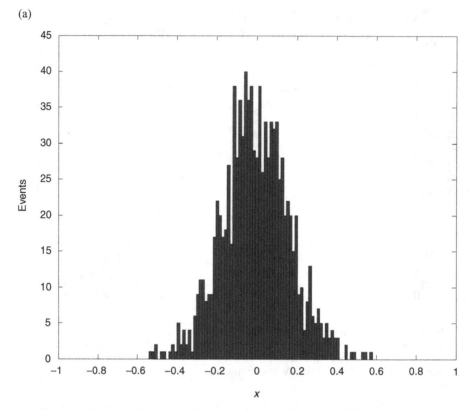

Figure 3.29   Simple Monte Carlo results for two body gluon–gluon scattering at a
mass of 200 GeV and a CM energy of 2 TeV. (a) Distribution of the momentum fraction, x, of the produced state. (b) Distribution of rapidity, y_3, of one of the final states
gluons.

A simple Monte Carlo program was written to simulate two body gluon–gluon scattering at a fixed mass of 200 GeV at the Tevatron. Results of the model are shown in
Fig. 3.29. We see that the x, $x = x_1 - x_2$, distribution for the composite state of mass
M is sharply peaked around the value of zero. Values for x are limited to be about zero
by the falloff of the parton distribution functions at large x. The plateau for the "decay"
products exists and is limited to $\Delta y \sim 3$ at the Tevatron for this mass. This is a kinematic,
not a dynamic, effect.

As another example of a two body decay angular distribution we look at the production
of both a W boson and a photon. The COMPHEP Feynman diagrams for this process
are shown in Fig. 3.30. This is another specific example of the production of a pair
of electroweak gauge bosons. These processes depend on the triple coupling of gauge
bosons, in this case the $WW\gamma$ vertex.

The angular distribution for the W plus photon production process at the parton level
is shown in Fig. 3.31. Note the strong forward–backward peaking of the angular distribution. This is due to the virtual exchange of the u and d quarks, similar to that observed in

(b)

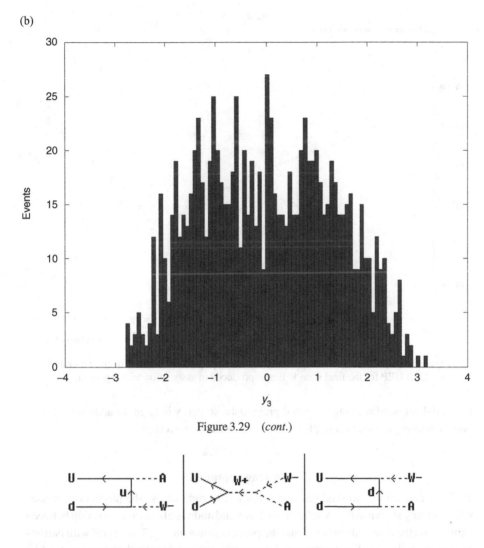

Figure 3.29   (*cont.*)

Figure 3.30   COMPHEP Feynman diagrams for the production of a pair of gauge bosons, a photon, and a *W* boson.

Rutherford scattering with the exchange of a photon. In addition, the angular distribution has a zero. This very distinctive SM prediction could be confirmed with a large enough event sample. Such a sample is not yet available at the Tevatron, although the process itself has been detected.

The scattering angle can be found using the measurements of the rapidities of the two partons in the final state, y_3 and y_4, as we show in Appendix C. The correlation between the rapidity of the final state particles in a simple Monte Carlo program with a fixed mass of 200 GeV for 2 TeV CM energy p–p collisions is shown in Fig. 3.32. Note the boundary illustrating the kinematic limit at large rapidity. There is, in addition, a strong

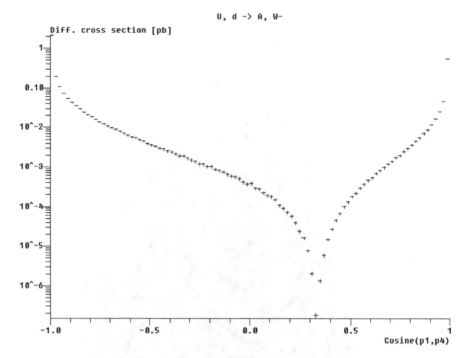

Figure 3.31   Distribution of the W angle with respect to the initial u quark generated by COMPHEP for the final state W boson produced in association with a photon.

forward–backward peaking, as noted previously, so that y is large. In addition, the two body scattering correlation implies $y_3 \sim -y_4$, at least on average.

3.7 Jet fragmentation

We have now almost worked our way from left to right across the physical processes schematically shown in Fig. 3.4. So far we have said nothing about the distinction between partons and the detected particles, and the process shown in Fig. 3.4 cuts off with partons exiting the collision. For fundamental, approximately stable, final state particles like electrons, photons, and muons there is really no distinction as these particles themselves are detected. For the quarks, gluons, and neutrinos we really need to look at the jets and not the partons. We have thus far simply used quark and gluon interchangeably with jet.

The Monte Carlo modeling of the parton to jet "fragmentation" is done in a series of complex programs that are available to researchers in this specialized area. For this text, COMPHEP evaluates the distribution functions, $f(x)$, properly, and the Standard Model dynamics, but not the fragmentation. We can also write our own simple Monte Carlo programs to crudely simulate the fragmentation of quarks/gluons into jets, and this has been done for the purposes of this text. In general, we will not focus on these experimental details here but will rather stick to the fundamental physics. Interested readers can find and execute PYTHIA, HERWIG, ISAJET, or some other of these complex computer

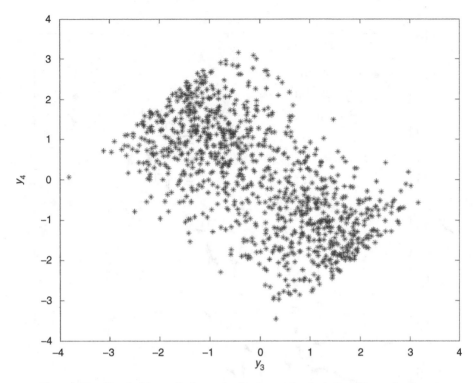

Figure 3.32 Simple Monte Carlo results for the production of a gluon pair of mass 200 GeV at a CM energy of 2 TeV in p–p collisions. The scatter plot shows the correlation between the rapidities of the two final state gluons.

codes. For example PYTHIA, a very popular program in high energy physics circles, is described in [11].

Some experimental data on fragmentation from both electron–positron annihilations and p–(anti)p collisions are shown in Fig. 3.33. For the pion fragments of jets there is shown a distribution in the momentum fraction, z, of the parent momentum, P, taken off by the pion of momentum, k. It is roughly independent of the energy of the parent for $z > 0.1$ and falls rapidly with increasing z. In addition, the multiplicity of charged fragments grows, on average, as the logarithm of the CM energy.

The fragmentation of quarks and gluons has already been introduced in Chapter 2. Fragmentation properties are assumed to "factorize" so that the way in which a parent quark or gluon fragments is independent of the mechanism by which the parent is created. Therefore we need only a single unified description of the fragmentation or "hadronization" process.

We assume for simplicity that all fragments are pions. We also assume that the transverse momentum acquired in the fragmentation process is limited with the fragment momentum transverse to the parent jet axis, k_T, limited to a value $\sim \Lambda_{QCD}$. The fragmentation function, $D(z)$, describes the distribution in $z = k/P$ of those products, where

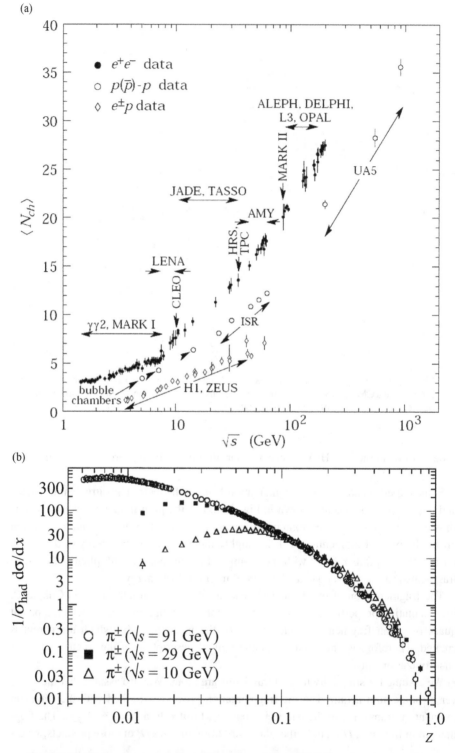

Figure 3.33   Fragmentation of a jet in electron–positron annihilations into an ensemble of final state hadrons. (a) Multiplicity of charged hadrons as a function of the energy of the $e^+ e^-$, p–$(\bar{p})p$, ep initial states. (b) Momentum fraction of the produced pions with respect to the initial electron momentum ([3] – with permission).

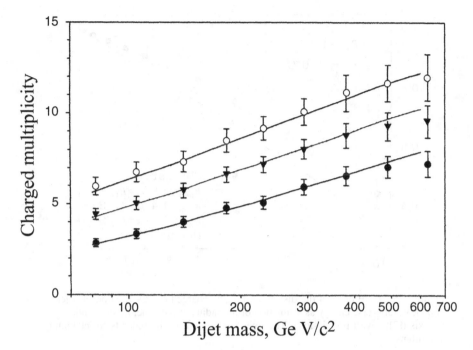

Figure 3.34 CDF data on the mean multiplicity of charged particles within a jet as a function of the mass of the jet–jet system. Note the semi-logarithmic scale. Data for different cone sizes about the jet axis (closed circle, $R = 0.17$; triangle, $R = 0.28$; open circle, $R = 0.47$) are shown ([12] – with permission). The lines are Monte Carlo predictions.

z is the momentum fraction of the parent, momentum P, carried off by the fragment, momentum k, $z_{min} < z < 1$, $z_{min} = M_\pi/P$. It has a "radiative form" similar to that already assumed for the parton distribution functions. This assumed form leads to a jet multiplicity, n, which is logarithmic in P in agreement with the data shown in Fig. 3.33.

$$z D(z) = a(1 - z)^\alpha,$$
$$\langle n \rangle = \int D(z) dz \sim a \int_{M/P}^1 dz/z \sim a \ln(P/M_\pi). \tag{3.17}$$

The fragmentation process implies that we observe a "jet" of particles, which move approximately along the direction of the parent quark or gluon. We expect a "core" within the jet which carries most of the jet momentum and which is localized at small cone radius, R, in (η, ϕ) space with respect to the jet axis. The core is surrounded at larger R by many low energy particles.

Data from CDF on the jet charged multiplicity are shown in Fig. 3.34 as a function of the mass of the jet–jet (or dijet) system. Note the expected logarithmic dependence of the mean charged particle multiplicity on the dijet mass. The existence of a sharply

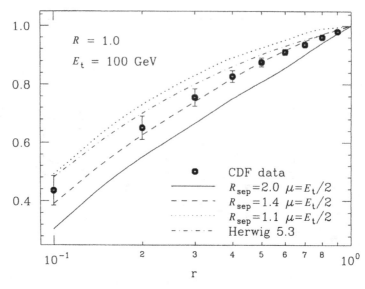

Figure 3.35   CDF data on the distribution of the charged energy fraction of a jet of 100 GeV transverse energy as a function of the radius of the cone, R, surrounding the jet axis ([7] – with permission). The lines correspond to different fits to jet finding algorithms.

peaked distribution of particles about the jet axis is also very evident because data are presented for different "cone" half angles.

More detailed information on the energy flow within a jet as a function of the cone radius R is shown in Fig. 3.35. As we can see, 40% of the energy of the jet is contained in a cone of radius $R = 0.1$ (Chapter 2 defines R as the radius in pseudorapidity–azimuthal angle space, $R = \sqrt{\Delta\eta^2 + \Delta\phi^2}$), while 80% is contained in a cone with radius $R = 0.4$.

These data can be compared with that derived from a simple Monte Carlo program that was written to model jet fragmentation. In the model, a series of massless fragments is picked out of a simple $D(z)$ distribution and they are then assigned a transverse momentum from a distribution similar to that shown in Fig. 3.5. This particular model uses $zD(z) \sim (1-z)^5$ and $\langle k_T \rangle \sim 0.72$ GeV. The "leading fragment," or particle with the highest z in the jet, is expected in this model to have $\langle z_{max} \rangle \sim 0.23$. Hence, on average the highest energy pion in a jet takes about $1/4$ of the jet momentum in this model. Results are shown in Fig. 3.36. There is qualitative agreement in the R dependence in the two figures, but not agreement in detail.

We must resort to experimental data because fragmentation is soft and thus non-perturbative, as was the case for the distribution functions of partons found in the proton. Clearly, the simple model and the data are in some rough agreement. However, a serious comparison with data requires a much more sophisticated treatment including final state parton showers and initial state gluon radiation. We will, in general, evade these

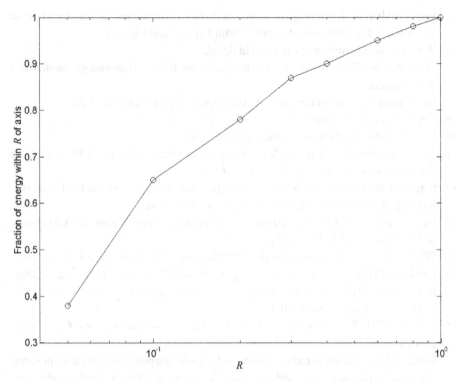

Figure 3.36 Simple Monte Carlo model corresponding to the data shown in Fig. 3.35 for comparison. The energy fraction of the jet fragments found within a cone of variable radius R centered on the jet axis is plotted vs. R.

complications and assume that an understanding of the physics of the processes is of more interest than a detailed, but purely phenomenological treatment of fragmentation.

Now we are armed with the ability to first estimate and then calculate all processes that exist in the Standard Model using simple formulae and then COMPHEP. We will apply our tools in Chapter 4 to data taken at the Tevatron. This data presently defines the "state of the art" for physics in proton–(anti)proton collisions. What is of crucial importance is that the student has the necessary tools. He or she can then duplicate most of the material given in this text.

Exercises

1. Use Eq. (C.2) to show that $y = \sinh^{-1}(P_\parallel / \sqrt{P_T^2 + M^2})$.
2. Show that y is additive under Lorentz transformation.
3. Show that y is approximated by pseudorapidity for zero mass particles.
4. Use the result of Exercise 1 to derive Eq. (3.1).

5. Run COMPHEP for $g + g \rightarrow g + g$ at 2 TeV. Plot the rapidity and transverse momentum distributions and compare with Fig. 3.2 and Fig. 3.3.

6. Work out the derivation of Eq. (3.4) in detail.

7. Show that the "Cerenkov" relationship, $\omega = k \cos \theta$, follows from energy–momentum conservation.

8. For b quark pair production at the LHC, estimate $xg(x)$ using Eq. (3.7).

9. Derive Eq. (3.8) in detail.

10. Show that the Jacobean is as stated, $dx_1 \, dx_2 = d\tau \, dy$.

11. Assuming that the decay width of the η_c charmonium state is 13 MeV, find the Drell–Yan cross section for $M \sim 3\mathrm{GeV}$, $J = 0$.

12. Establish the relationship between the initial state x values and the final state two body rapidities given in Eq. (3.17) (see Appendix C first).

13. For a pion mass of 0.14 GeV, estimate the mean multiplicity of pions at a CM energy of 1 TeV for $a \sim 3$ in Eq. (3.17).

14. What is the average emission angle of the leading jet fragment for a 100 GeV jet?

15. Use COMPHEP to study $g + g \rightarrow g + g$ at 100 GeV CM energy. Is the result stable? If not, why? Try a cut on the final gluon transverse momenta of > 10 GeV. Is this more stable? (See Appendix B.)

16. Use COMPHEP to compare $u + \bar{u} \rightarrow Z$ and $d + \bar{d} \rightarrow Z$ at the same CM energy. Can you explain the ratio of the cross section?

17. Use COMPHEP to study radiated photons. Consider the process of electron–positron elastic scattering with a radiated photon at CM energy of 100 GeV. Look at the energy of the photon and the angle with respect to the incident electron. Are the photons soft and collinear?

18. Use COMPHEP to study the angular distribution in $u, U \rightarrow e1, E1$. Look at the cosine of the angle between particles, 1 and 3 (u quark and electron) at 50, 90, and 150 GeV. How does the asymmetry change with CM energy?

19. Do the same for proton–antiproton scattering as for partons in Exercise 18. Compare with the Monte Carlo results presented in the text.

References

1. Blazey, G. and Flaugher, B., Fermilab-Pub-99/038.
2. CDF Collaboration, *Phys. Rev. Lett.*, **61**, 1819 (1988).
3. Particle Data Group, *Review of Particle Properties* (2003).
4. Barger, V. and Phillips, R., *Collider Physics*, Redwood City, CA, Addison-Wesley Publishing Company (1987).
5. Bauer, U., Ellis, R. K., and Zeppenfeld, D., Fermilab-Pub-00/297 (2000).
6. CDF Collaboration, arXiv:hep-ex/0106047 (2001).
7. Huth, J. and Mangano, M., *Ann. Rev. Nucl. Part. Sci.*, **43**, 585 (1993).
8. D0 Collaboration, Fermilab-Pub-00/213-E (2000).
9. D0 Collaboration, *Phys. Rev. Lett.*, **87**, 251805-1 (2001).
10. CDF Collaboration, Fermilab-Pub-00/133-E (2000).
11. Sjostrand, T., *Phys. Lett. 157B*, **321** (1985).
12. CDF Collaboration, Fermilab-Pub-01/106-E (2001).

Further reading

Altarelli, G. and L. Dilella, *Proton–Antiproton Collider Physics*, Singapore, World Scientific (1989).

Barger, V. D. and R. J. N. Phillips, *Collider Physics*, New York, Addison-Wesley (1987).

Dilella, L., Jet production in hadronic collisions, *Ann. Rev. Nucl. Part. Sci.* **35**, 107 (1985).

Eichten, E., I. Hinchliffe, K. Lane, and C. Quigg, *Supercollider Physics*, *Rev. Mod. Phys.* **56**, 4 (1984).

Feynman, R. P., *Photon–Hadron Interactions*, Reading, Massachusetts, Benjamin (1972).

James, F., Monte Carlo theory and practice, *Rep. Progr. Phys.*, **43**, 1145 (1980).

Owens, J., Large momentum transfer production of direct photons, jets and particles, *Rev. Mod. Phys.* **59**, 465 (1987).

Owens, J. and W. Tung, Parton distribution functions of hadrons, *Ann. Rev. Nucl. Part. Sci.* (1992).

4

Tevatron physics

True science teaches, above all, to doubt, and to be ignorant.

Miguel de Unamuno

Rules and Models destroy genius and art.

William Hazlitt

We have now obtained the tools we need to examine the production of SM particles in p–(anti)p collisions. In this chapter, our aim is to see where the frontier of this knowledge presently is. The Tevatron accelerator complex operated at the Fermi National Accelerator Laboratory (Fermilab or FNAL) has the highest available CM energy of 1.96 TeV. There are two general purpose experiments taking data at Fermilab, CDF and D0. Some data taken by these experiments have already been shown in previous chapters, and we will examine more now. They will define, for high transverse momentum processes, what we know and how we know it.

The statistical power of these data will be improved because CDF and D0 resumed data taking in 2001. The rate increase should allow studies of gauge boson pairs and searches for low mass Higgs particles at the Tevatron. In 2007 the Large Hadron Collider (LHC) operated at the European high energy facilities at CERN will begin operations at a CM energy of 14 TeV.

4.1 QCD – jets and dijets

One of the processes with the largest cross section is jet production because it is a strong interaction process and because the gluons are the dominant parton in the proton at low x values. The simplest measurement is the distribution of transverse energy for any produced jet or "inclusive jet" E_T. A jet is defined experimentally as localized energy in a cone of radius R, with $R \sim 0.5$. Data from D0 for this process are shown in Fig. 4.1. Note the rapid falloff with increasing transverse momentum. Clearly, the QCD theory works well and fits the data over many orders of magnitude.

The data extend out to a substantial fraction of the kinematic limit which occurs when the dijet mass $M = \sqrt{s}$, $E_T = \sqrt{s}/2 \sim 900$ GeV. Historically, particle scattering at wide angles has led to the discovery of substructure. The most well known example is in Rutherford scattering, where the existence of large angle scattering events led to the hypothesis of an atomic structure with widely distributed electrons and the very localized nucleus as a substructure. More recently, wide-angle scatters of leptons from protons

(a)

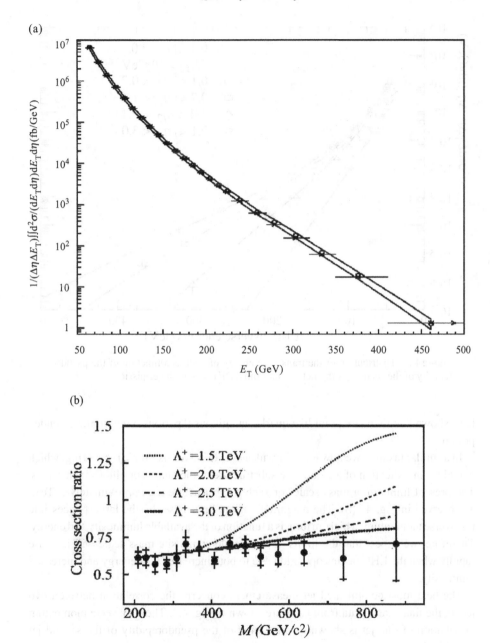

(b)

Figure 4.1   Jet production at the Tevatron in D0 ([1] – with permission). (a) Distribution of the transverse momentum of a single jet. (b) Limits on the energy scale for quark "compositeness". The solid lines are SM Monte Carlo predictions.

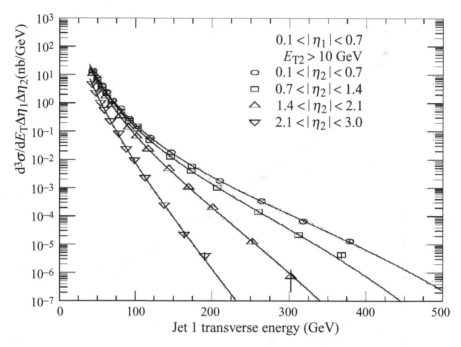

Figure 4.2   Distribution of the transverse energy of a jet as a function of the pseudorapidity of the second jet in dijet events from CDF ([2] – with permission).

have shown that there are point like quarks and gluons ("partons") within the extended proton.

In a similar fashion, we now look for wide angle (*S* wave – isotropic) scattering, which would be an indication of a composite substructure of the quarks or gluons themselves. The present limits on a mass scale for such a substructure are approximately 2 TeV, as observed in Fig. 4.1(b). The magnitude of the limit is set by the largest accessible transverse momentum, which, in turn, is a function of the available luminosity and energy. Therefore, we expect that the limits on a possible composite mass scale will increase rapidly when the LHC begins operation due to both increased CM energy and increased luminosity.

The next most complicated jet measurement concerns the correlation between two jets in the final state. Data from CDF are shown in Fig. 4.2. The transverse momentum distribution of one jet is shown as a function of the pseudorapidity of the second jet found in the event.

Clearly, as the magnitude of the pseudorapidity difference of the final state jets, $|\eta_3 - \eta_4|$, increases, M_{34} increases (see Chapter 3) and the cross section decreases at least as rapidly as the third power of the mass. As we can see, QCD also describes the dijet data very well over a wide range of cross section values.

Next we look at the mass distribution of the dijets. Data from the D0 experiment are shown in Fig. 4.3. We expect that the distribution falls with a $1/M^3$

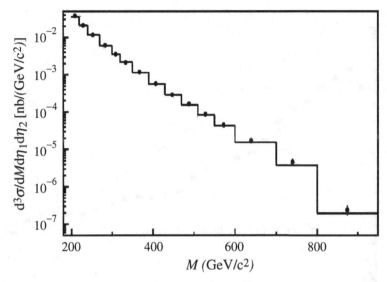

Figure 4.3   Distribution of the mass of dijets from D0 for jets produced at low rapidities ([3] – with permission). The line is a Monte Carlo prediction.

behavior due to the underlying point like parton scattering and contains a second factor $\sim (1 - M/\sqrt{s})^{12}$ due to the gluon initial state distribution functions. As with the transverse momentum distribution, we can look for anomalous production of high mass dijets as possible evidence for quark or gluon compositeness. However, as seen in Fig. 4.3, QCD appears to explain the data well out to jet–jet masses of $\sim$0.8 TeV.

The limit that we can place on the resonant production of possible excited quarks is shown in Fig. 4.4. The process of producing excited quark states is imagined to be that quarks and gluons would form a resonant state at the mass of the excited quark similar to the Drell–Yan mechanism we studied in Chapter 3, $q + g \rightarrow q^* \rightarrow q + g$, where q^* indicates the resonant excited quark state. The absence of such resonant structure in the mass distribution allows D0 to set a limit on the mass of such states of 725 GeV. Up to this mass the quarks act like fundamental point like particles containing no internal states that can be excited.

The dijet angular distribution has also been published. Data from the D0 experiment are shown in Fig. 4.5 for different dijet mass intervals.

If gluon exchange describes the dynamics of jet–jet production (see Chapter 3 for the g–g Feynman diagrams), then the distribution of the variable χ, where $\chi = (1 + \cos \hat{\theta}) / (1 - \cos \hat{\theta})$, is flat, $d\hat{\sigma}/d\chi \sim$ constant, as it is in the familiar case of Rutherford scattering. Recall that (Appendix C) the scattering angle can be determined from measurements of the transverse momentum and the rapidity of the two jets in the final state. The variable $\hat{t}$ is the square of the parton momentum transfer, which in the reaction $1 + 2 \rightarrow 3 + 4$ is $(p_1 - p_3)_\mu \cdot (p_1 - p_3)^\mu = -2\,\hat{p}^2\,(1 - \cos \hat{\theta})$ for massless partons. The effect of the exchanged gluon propagator on the differential cross section is removed by the change

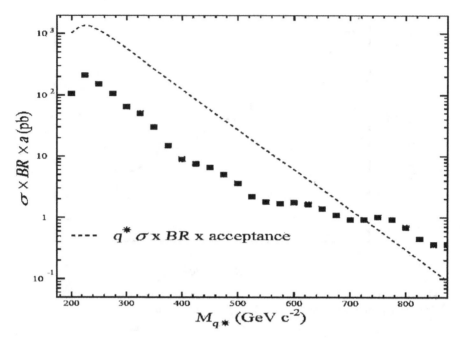

Figure 4.4   The cross section for the production of excited quarks as a function of the mass of the excited quark. The lack of resonant structure in the mass distribution, Fig. 4.3, leads to a limit on the cross section at each mass (dashed curve) and thus on the mass of the excited quark of >725 GeV, which is the mass where the cross section limit equals the production cross section for excited quarks ([3] – with permission).

of variable, $\hat{t} \rightarrow \chi$.

$$d\chi \, d\hat{t} \sim 1\hat{t}^2. \tag{4.1}$$

The results for small angles are particularly simple, $\chi \rightarrow 4 \, / \, \hat{\theta}^2, \hat{t} \rightarrow (\hat{p}\hat{\theta})^2, \chi \rightarrow (2\hat{p})^2 \, / \, \hat{t}$. We expect that point like scattering describes the fundamental $2 \rightarrow 2$ process. Therefore, we expect that the χ distribution is uniform. There are small higher order corrections to the distributions that are evident in Fig. 4.5 and that are calculable. Since there are no deviations from the QCD theoretical distributions at large scattering angles, we conclude that there is no evidence for the existence of composite quarks at this time.

4.2 α_s determination

In quantum field theory the coupling "constants" of the three SM forces that appear in the Lagrangian have "effective" values, which are functions of the mass scale at which they are examined. This effect is due to quantum corrections caused by higher order diagrams, as discussed in some detail in Appendix D. We can use existing jet data to validate the QCD prediction for the change of α_s with the mass scale Q.

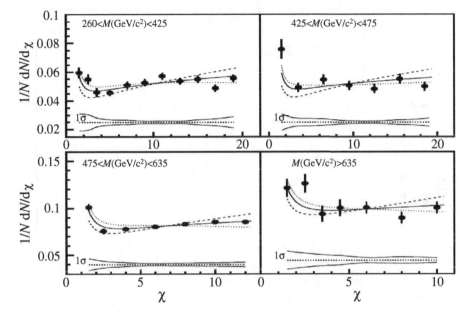

Figure 4.5 Distribution of the scattering angle variable χ for different values of the jet–jet invariant mass obtained by the D0 experiment. The curves represent the predictions of QCD perturbation theory ([2] – with permission). The lower set of curves shows the estimated systematic errors.

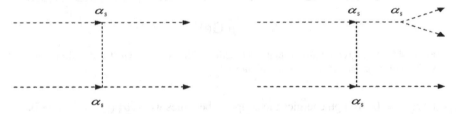

Figure 4.6 Schematic representation of the scattering due to the mutual interaction of gluons, $g + g \to g + g$, $g + g + g$. Note the triple gluon fundamental vertex that exists in QCD.

In QCD the gluons mutually interact because they, themselves, carry "color." This is illustrated very schematically in Fig. 4.6. Roughly speaking, the ratio of three jets to two jets in the final state is given by the strength of the strong coupling constant (see Fig. 4.6). That ratio can then be studied experimentally as a function of the mass scale of the jet events. In that way we can measure experimentally how the coupling constants "run" with mass scale.

The mutual self-coupling of gluons leads to the conclusion that the strong coupling strength actually decreases as the mass increases, opposite to the behavior of electromagnetic charge. On the other hand, the coupling becomes very strong at large distance scales. For QCD we define an energy scale Λ_{QCD} where the interactions become strong,

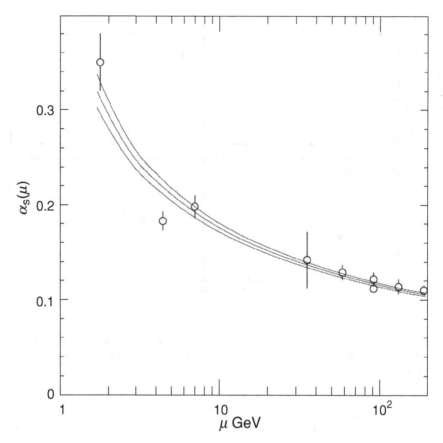

Figure 4.7   Strong coupling constant as a function of the mass scale μ. The data decrease with increasing mass ([4] – with permission).

$1/\alpha_s(\Lambda_{QCD}^2) = 0$. At high energies the coupling becomes weak, $\alpha_s\,(Q^2 \to \infty) \to 0$.

$$\alpha_s(Q^2) = [12\pi/(33 - 2n_f)]/\ln\!\big(Q^2/\Lambda_{QCD}^2\big)]. \qquad (4.2)$$

In Eq. (4.2) n_f is the number of fermion generations that are "active", or have a mass below the mass scale Q. For example, we give numerical values at a few mass scales. We take the QCD mass scale to be $\Lambda_{QCD} \sim 0.20\,\text{GeV} \sim 1$ fm. At the Z mass, the strong interactions are appreciably weaker than at the $\sim$GeV mass scale.

$$\alpha_s((1\ \text{GeV})^2) = 0.55,$$
$$\alpha_s((10\ \text{GeV})^2) = 0.23,$$
$$\alpha_s\!\big(M_Z{}^2\big) = 0.15. \qquad (4.3)$$

Experimental data on the strong coupling constant as a function of mass scale are shown in Fig. 4.7. Note the rapid falloff from the 0.2 GeV scale where the strong interactions are strong. The scale for Q is logarithmic.

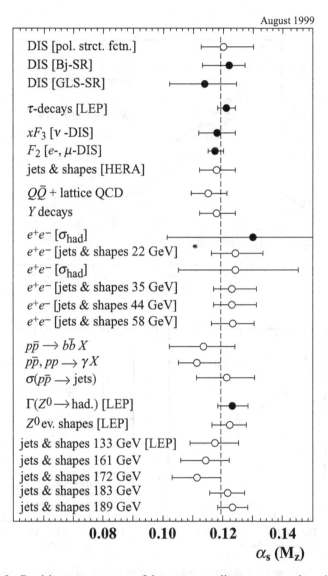

Figure 4.8 Precision measurements of the strong coupling constant evaluated at the Z mass. Data come from measurements of lepton–*p* scattering, electron–positron production of jets, and proton–(anti)proton jet production as well as other reactions ([5] – with permission).

A collection of precision measurements of the strong coupling constant extrapolated to the Z mass is given in Fig. 4.8. Many of these measurements come from data on the production of jets, either at proton–(anti)proton colliders or at electron–positron colliders. The data appear to have converged to a value for the strong coupling constant of roughly 0.12 at the Z mass.

(a)

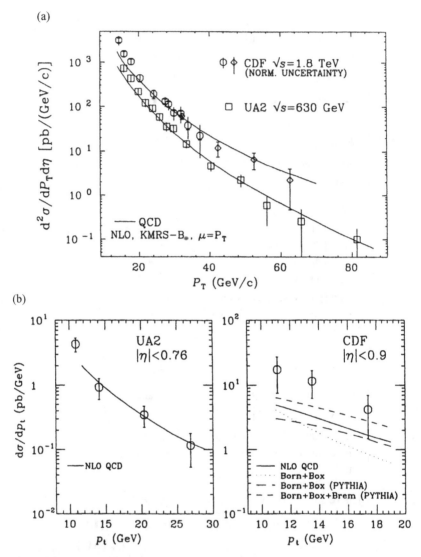

(b)

Figure 4.9 Data on the transverse momentum distribution of (a) single photon and (b) di-photon production ([6] – with permission) at CDF and UA2 (the UA1 and UA2 experiments were operated at a proton–antiproton collider located at CERN with collisions of 0.63 TeV energy in the CM).

4.3 Prompt photons

Now let us generalize slightly from gluon jets to the study of reactions with a single photon or two photons in the final state. Data from CDF and the CERN experiment UA2 are shown in Fig. 4.9. The distribution of transverse momentum of a final state photon is shown. The smaller value of the cross section with respect to jets limits the statistical power of the data, and hence the transverse momenta are limited to fairly low

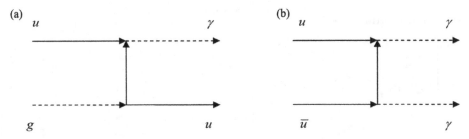

Figure 4.10   Schematic representation of the Feynman diagrams in the Born approximation for the production of (a) single photons and (b) di-photons.

values. However, the data are in reasonably good agreement with the SM prediction except perhaps at very low transverse momentum. It may be true that an "intrinsic" parton transverse momentum of ~ 3 GeV or less is necessary to explain the data. As we noted before, the quarks are bound in the proton, so that some transverse momentum ~ 0.2 GeV is expected.

As already mentioned in Chapter 3, these are two-to-two processes with kinematic relationships similar to those found in jet production. The dynamics of the fundamental point like parton scattering are also similar. The cross section level is reduced with respect to gluon–gluon scattering by coupling strengths and initial state parton source factors. A schematic representation of the lowest order diagrams for single and double photon production is given in Fig. 4.10. Clearly, two photon production in the Born approximation shown here is just another generic two-to-two process.

These data for single photons plus jets can be used in "jet balancing" for the calibration of hadron calorimeters, as we mentioned in Chapter 2. It is easier to balance a photon–jet than a jet–jet event because precision electromagnetic calorimetry (see Chapter 2) can be used to accurately measure the photon and then predict the jet energy, while jet energy measurements have intrinsic fluctuations (see Chapter 2) and thus have worse energy resolution.

The two-photon process constitutes an important SM background in Higgs searches. Therefore, it is important to insure that we have a good understanding of this background so that we can extrapolate to the LHC. COMPHEP Monte Carlo predictions for the transverse momentum distribution of the photon are shown in Fig. 4.11. New data from CDF and D0 at higher photon transverse momenta will be important in comparing with the Monte Carlo predictions. The COMPHEP program does not include higher level processes such as internal "loops" or "box" diagrams, which may be important in two-photon production. The COMPHEP user must be aware of the limitations of this program in comparing with real data.

The data from CDF, Fig. 4.9, are a factor of three above the COMPHEP predictions at photon transverse momenta ~ 10 GeV, where CDF finds a cross section ~ 30 pb/GeV. "Intrinsic" parton momentum is one mechanism that has been postulated in order to improve the agreement of the model with the data since it serves to smear the sharp

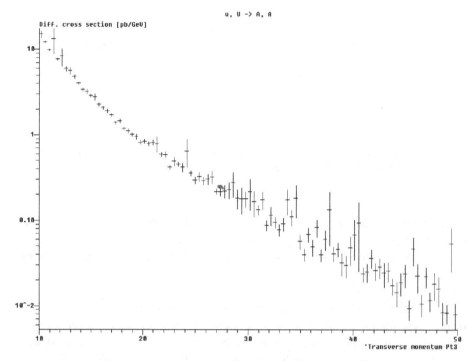

Figure 4.11   COMPHEP results for the cross section as a function of the transverse momentum of one of the photons in di-photon production in p–p collisions at 2 TeV CM energy.

falloff of the cross section with transverse momentum. More data are needed before a firm conclusion on the existence of "intrinsic" momenta of a few GeV can be made.

4.4 *b* production at FNAL

In this book, we will say little about the production of c and b quarks and their subsequent use in the study of the basic properties of weak decays of quarks. We are interested in physics at the energy frontier, which means we are concerned with the highest available mass scales. Indeed, there are many fine textbooks written solely about B physics. There are accelerators and associated detectors dedicated to B physics studies operating in Japan, at SLAC, and at Cornell in the US. Clearly, this is a field of study that well deserves a volume by itself.

Nevertheless, we briefly show here some Tevatron data on the production of states containing b quarks. We do this because many of the Higgs and new phenomena search strategies rely on the identification of hadrons containing b quarks in the final state. Therefore, the background processes must be well understood if an incisive search is to be made. In Fig. 4.12 we show the transverse momentum distribution of b quarks produced in the CDF and D0 experiments. The natural mass scale for the production is that of the quark mass itself. Because the b mass is about 5 GeV, we expect that

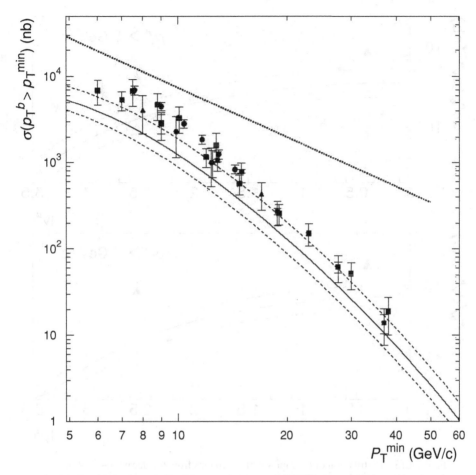

Figure 4.12 Data from CDF and D0 on the cross section for the production of *b* quarks as a function of the minimum transverse momentum of the *b* quark, $P_T{}^{min}$. The dotted line is drawn to indicate the expected behavior of the cross section ([7] – with permission). The solid line and accompanying dotted lines indicate a Monte Carlo model fit and the associated theoretical uncertainties.

perturbative QCD should work properly, since $m_b \gg \Lambda_{QCD}$, which implies $\alpha_s (m_b) \ll 1$ (see Fig. 4.7).

The expected Rutherford like behavior for two body scattering is $d\sigma/dP_T \sim 1/P_T^3$ so that $\sigma(P_T > P_{T\,min}) \sim 1/P_{T\,min}^2$. This behavior roughly corresponds to the data at low transverse momenta where the effects of the falloff of the parton distribution functions with *x* are not expected to be important.

In Fig. 4.13 we show D0 data on the rapidity distribution of muons arising from the decays of *B* mesons. There is an evident rapidity "plateau" which extends to $y_{max} \sim 2.5$ as expected at this low mass ($\sim 2m_b$) scale. These data are in rough agreement for the shape of the rapidity distribution with the Monte Carlo predictions available to the researchers, which are the curves shown in Figs. 4.12 and 4.13. However, the agreement is not good,

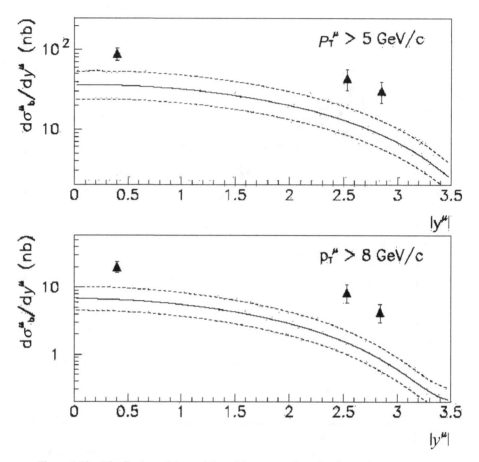

Figure 4.13   Distribution of the rapidity of the muons from the decay, $b \to c + \mu^- + \bar{\nu}_\mu$, of the b quarks for two different minimum requirements on the muon transverse momentum ([1] – with permission). The lines correspond to theoretical predictions and their uncertainty.

which means that background calculations for new phenomena such as Higgs decays into b quark pairs should also be assumed not to be terribly reliable.

The CDF experiment has taken high quality data on the lifetimes of states containing b quarks. As mentioned in Chapter 2, a silicon vertex tracker is used to find decay vertices and the production vertex. The decay distance and the momentum of the reconstructed B particle decay allow CDF to make precision lifetime measurements. The B particles are typically bound states of quarks and antiquarks, which contain b quarks.

Prior to the advent of high quality silicon vertex tracking detectors, the study of B decays was hampered in p–(anti)p colliders by the many confusing background tracks that exist in the "underlying event". Thus, silicon detectors were the enabling technology for p–(anti)p colliders in the study of the decays of heavy quarks. Data are shown from these analyses in Fig. 4.14. The Λ_b^0 is a three quark bound state (*bud*) like the neutron (*dud*).

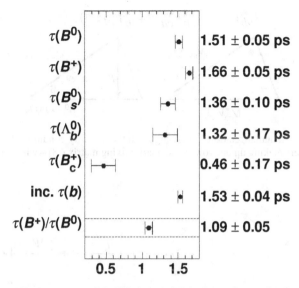

Figure 4.14 Measurements of the lifetimes of *B* hadrons (in ps) for those colorless states containing *b* quarks from CDF ([1] – with permission).

Obviously, *p–p* collider experiments can make an impact on *b* physics research even though many competing *B* "factories" are operating at electron–positron colliders. The mean value for the lifetimes of all the states studied that contain a *b* quark is roughly 1.5 ps. The distance associated with that lifetime is about 450 μm, as quoted in Chapter 2. The exception shown in Fig. 4.14 has to do with the bound state of a *c* quark and a *b* antiquark, $B_c^+ = c\bar{b}$. Since either heavy quark can decay weakly, the lifetime is shorter than the lifetime of states containing only *b* quarks and light quarks, $\Gamma \sim \Gamma_b + \Gamma_c$, $\tau = 1/\Gamma < \tau_b$.

The mass scale for *b* quark production is $2m_b \sim 10$ GeV, so the cross section is large at the LHC, $\sim (0.1 - 1.0)$ mb, and the Tevatron. Therefore high statistics data can be obtained at hadron collider experiments as well as at e^+e^- machines such as Belle (Japan) and BaBar (SLAC). The large value of the *b* cross section opens up the possibility of high statistics studies of *b* quarks and searches for rare *b* decays. In addition, all the states shown in Fig. 4.14 are produced simultaneously, which is not the case in electron–positron colliders where the initial state CM energy is the same as the parton–parton CM energy since leptons are fundamental particles, while protons are not.

In the SM, all CP violation is due to a single complex phase of the mixing matrix $V_{qq'}$. A vigorous current area of research is to explore whether this SM assumption is found to be true in Nature. In Fig. 4.15 we show a schematic representation of one of the unitarity relationships for the weak mixing matrix, $V_{qq'}$, governing quarks decays, which are defined in Appendix A. Unitarity guarantees that the gauge coupling is of universal strength and implies that there are three and only three light generations of quarks and leptons. Initial results from BaBar and Belle indicate that the relationship shown in Fig. 4.15 is satisfied to the present experimental accuracy. Therefore, there is

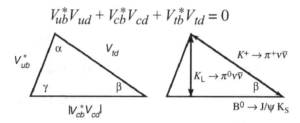

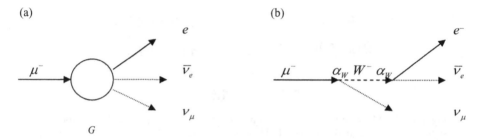

Figure 4.15   Triangular relationship arising from the unitarity conditions for the complex parameters making up one row of the quark mixing matrix ([8] – with permission).

(a)

(b)

Figure 4.16   Schematic representation of the decay of the muon. (a) Representation of the muon decay as an effective four fermion interaction described by the Fermi coupling constant G. (b) The same decay viewed as the virtual emission of a W gauge boson followed by its virtual two-body decay into a lepton pair.

no indication as yet for additional CP violation effects due to, for example, SUSY (see Chapter 6). Ever more precise data taken concerning B and K decays continue to refine the over constrained unitarity relations.

4.5 t production at Fermilab

The top quark has a mass of 175 GeV, as determined by direct measurement at Fermilab. Because the mass is so high, top quark events have only been produced and studied at the Tevatron.

Of all the quarks and leptons, only the top quark has kinematically allowed two body weak decays where a real W boson is produced rather than a virtual W. The decay is $t \rightarrow W^+ + b$. Schematic diagrams for muon three-body weak decays are shown in Fig. 4.16. At low mass scales, weak decays can be viewed as an effective four-fermion interaction, characterized by the Fermi coupling constant, G, which can be determined by measuring the rate for muon decay, $\mu^- \rightarrow e^- + \bar{\nu}_e + \nu_\mu$. Because $\Gamma \sim |A|^2 \sim G^2$ and $[G] = 1/M^2$, $[\Gamma] = M$, there must be five powers of the muon mass to make the dimensionally correct estimate, $\Gamma \sim G^2 m_\mu^5$.

At a more fundamental viewing (see Appendix A) muon decay can be thought of as the virtual emission of a W boson and a muon neutrino, with strength α_W in the decay width, followed by a propagator at low momentum transfer contributing a factor $1/M_W^2$ in the amplitude (see Eq. (1.6)), and ending with the subsequent virtual decay of the W into an electron and an electron antineutrino, contributing another factor of α_W to the decay rate: $\mu^- \to \nu_\mu + W^- \to \nu_\mu + (e^- + \bar{\nu}_e)$. The lifetime of the muon can again be estimated by dimensional arguments. It contains coupling factors due to the two weak vertices and the W propagator, as we mentioned in Chapter 2. However, this method gives a poor estimate because there is, unfortunately, a large, dimensionless, purely numerical factor, $1/[192\pi^3]$.

The correct expression for the muon lifetime is given in Eq. (4.4), along with the expression derived using dimensional arguments.

$$\Gamma_\mu = G^2 m_\mu^5 / 192\pi^3$$
$$\sim \alpha_W^2 (m_\mu / M_W)^4 m_\mu. \qquad (4.4)$$

For all quarks except the top, the decay width is much less than the strong binding energy scale $\sim \Lambda_{QCD}$. This fact implies that all quarks except the top form strong bound states before they decay, for example the b forms a $B^- = (b\,\bar{u})$ meson before it decays. For the top quark, the decay width is large, which tells us that the top decays before it can form bound states. Therefore, we see the top quark at a fundamental level without the obscuring strong interactions.

For the top quark, we have available a direct two body decay with a single weak vertex, $t \to b + W^+$. In fact, the decay width of the top quark is quite comparable with the width of the W boson because both are direct two body decays. The top decays occur so rapidly ($\Gamma_t \gg \Lambda_{QCD}$) that no strongly bound top–antitop bound states are formed as they are for c (charmonium) and b (bottomonium) quark–antiquark pairs. The expression for the top width, Eq. (4.5), is first order in the Fermi constant G due to the single vertex in the decay amplitude. Thus, since $[\Gamma] = M$, $[G] = M^{-2}$, we expect $\Gamma_t \sim G m_t^3$ or, alternatively, with $\alpha_W \sim G M_W^2$,

$$\Gamma_t = G m_t^3 / 8\pi \sqrt{2}$$
$$= (\alpha_W / 16)(m_t / M_W)^2 m_t \sim 1.76 \, \text{GeV}. \qquad (4.5)$$

Clearly, the ratio of decay widths for the top and a heavy quark, Q, is generically $\Gamma_Q / \Gamma_t \sim \alpha_W [m_Q^5 / m_t^3 M_W^2]$. Even for the bottom quark that ratio is much less than one, which indicates the uniqueness of the top among all the quarks.

Data from the D0 experiment on the spectroscopy of top quarks are shown in Fig. 4.17. Each produced top in the top pair final state almost always decays into a W boson and a b quark. In turn, the W can decay into a charged lepton and a neutrino or a quark–antiquark pair. The D0 data shown here use the lepton + jets final state, $t \to W + b$, $W \to J + J$, $\ell + \nu$, where one W decays into a lepton–neutrino while the other decays into a quark–antiquark pair or $t + \bar{t} \to (W + b) + (W + b) \to (\ell + \nu + b) + (J + J + b)$. Because of the neutrino in the final state the top mass is not particularly accurately reconstructed.

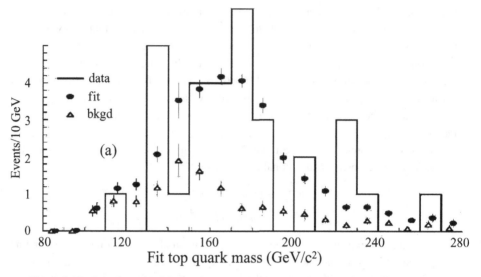

Figure 4.17 Data from the D0 experiment on the mass distribution of top quark candidates ([1] – with permission).

The final state is therefore a complex four jet + lepton + missing transverse energy event. In fact, early top candidate events for both CDF and D0 were already shown at the end of Chapter 2. As expected from our discussion in Chapter 2 on the accuracy of calorimetric mass reconstruction and missing transverse energy, the top quark experimental mass error is quite large. Nevertheless, although the experimental mass resolution is greater than the intrinsic width of the top, the mean value, or mass of the top, can still be determined very accurately.

Data from the CDF experiment on top production are shown in Fig. 4.18. The CDF detector was capable of b tagging (see Chapter 2) using the precision silicon inner "vertex" tracking. The data shown here have either one or two jets "tagged" as likely to be a heavy flavor jet as a requirement to accept potential candidates.

There are two b quarks in each top pair event, so b tagging capability is very important in reducing backgrounds from $W + W +$ four jet events. The CDF data shown here are for lepton + jets with silicon or b decay "lepton tags" (from $b \rightarrow c + \ell^- + \bar{\nu}_\ell$ decays). The plots are for no b tags or one or both b jets tagged in the silicon detector or by way of detection of a soft lepton tag (SLT) arising from leptonic b decays. The mass reconstruction of the top is again calorimetric. Therefore the intrinsic top width is again swamped by the instrumental resolution.

The data from CDF and D0 can be combined to form the world average for direct measurements at the Tevatron. A summary is shown in Fig. 4.19 for different final states corresponding to different decay modes of the W boson. The final state can be two b jets + two leptons + missing energy, four jets + one lepton + missing energy, or six jets. The combined data have an error on the top mass of about 5 GeV for data taken during the twentieth century. Future data taking beginning in 2001 with substantially increased

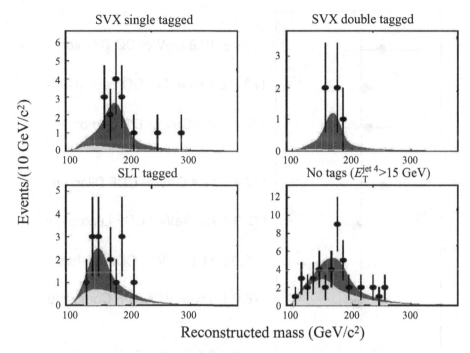

Figure 4.18   Data from the CDF experiment on the reconstructed mass distribution of top quark candidates. The four plots refer to data with different *b* quark "tagging" requirements. Expected distributions of both the top signal and residual backgrounds are indicated ([1] – with permission). The light shaded area is background, while the dark shading is the signal.

luminosity will considerably reduce this error. This improvement will, in turn, have an impact on the limits we can place on the Higgs mass (see Fig. 4.39).

The mass of the top quark is so large that we expect perturbative QCD to give a very good description of the production dynamics. Shown in Fig. 4.20 is the cross section for top pair production as a function of the top quark mass. There is good agreement with the measured production cross section at the measured top mass value. So far, there seems to be no mystery in the description of top production.

In Fig. 4.21 we show the COMPHEP Monte Carlo prediction for the gluon–gluon initiated cross section for top pair production as a function of the CM energy in p–p collisions. The gluon–gluon fundamental cross section, $g + g \rightarrow t + \bar{t}$, rises by a factor $\sim$600 in going from the Tevatron to the LHC. However, there are valence antiquarks available at the Tevatron, which softens this behavior somewhat, e.g. the process $u + \bar{u} \rightarrow t + \bar{t}$ is available at the Tevatron, but still a factor $\sim$100 rise in the cross section exists. That rise implies that strong top production is copious at the LHC. The resulting W pairs from top decay constitute a major background in some of the new particle searches, in addition to the rarer background from electroweak W pair production.

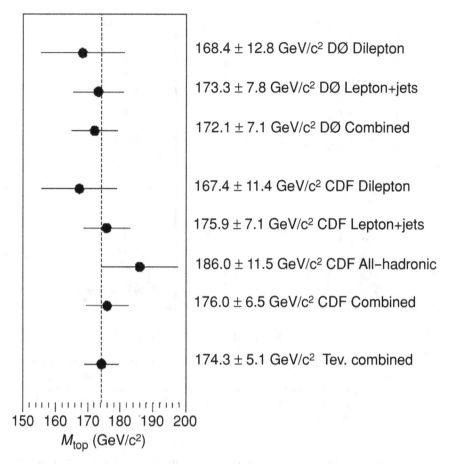

Figure 4.19 Data from both the D0 and CDF experiments on the measurement of the mass of the top quark ([1] – with permission).

4.6 DY and lepton composites

The cross section obtained by CDF for the production of dileptons as a function of their invariant mass is shown in Fig. 4.22. The fundamental process is the annihilation of quark and antiquark into a Z boson, $q + \bar{q} \to Z^0$, or a photon.

The annihilation of a quark and an antiquark in the initial state is called Drell–Yan production for historical reasons. The cross section falls rapidly as the mass increases. For a $\bar{u} + d$ initial state the W^- is a prominent feature of the spectrum, while for $\bar{u} + u$ initial states the Z is the main high mass feature. There is also a continuum from the reaction $u + \bar{u} \to \gamma^* \to \ell^+ + \ell^-$ caused by virtual photon production. Above the mass of the gauge bosons there is no known SM signal, and searches for new states beyond the SM such as "composite" leptons or heavy "sequential gauge bosons" recurring at higher masses are made by exploring the high mass part of the $\ell^- + \bar{\nu}_\ell$ or the $\ell^+ + \ell^-$ distributions. There is an observed "continuum" with a featureless background.

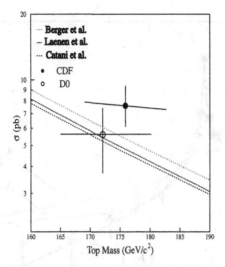

Figure 4.20   The predicted cross section for the production of the top quark as a function of the top quark mass at the Tevatron. Also shown is the top cross section measured by CDF and D0 ([9] – with permission). The different lines indicate a spread of theoretical predictions.

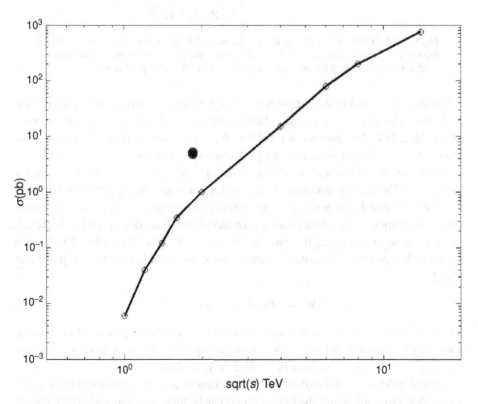

Figure 4.21   Cross section from COMPHEP for the production of top quark pairs as a function of the CM energy in *p–p* collisions (*g–g*). The dot indicates the Tevatron measurement in 1.8 TeV *p–p̄* collisions.

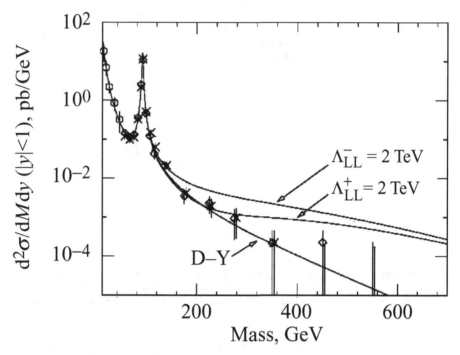

Figure 4.22   Data from CDF on the production of lepton pairs at high mass. The *Z* resonance is a prominent feature. The curves show the predicted anomalous production rates for composite leptons and the SM prediction ([10] – with permission).

The data shown in Fig. 4.22 show no unusual production of lepton–antilepton pairs at high mass. This allows CDF to place a limit on the mass scale for lepton "compositeness" of roughly 2 TeV. This limit is comparable to that assigned to quark compositeness mass scales set by the lack of anomalous jet production at high mass.

The data on the "transverse mass" distribution, M_T, of leptons and neutrinos are shown in Fig. 4.23. The transverse mass associated with a lepton and missing energy is defined in Eq. (4.6). Because the longitudinal component of missing energy is very poorly measured due to small angle energy disappearing unobserved (recall the discussion in Chapter 2), we are limited to measuring the mass in the transverse plane. Examples of individual events with produced *W* bosons have already been displayed in Chapter 2, Fig. 2.7 and Fig. 2.23.

$$M_T^2 = 2P_{Tl}|\not{E}_T|(1 - \cos\phi_{l|\not{E}_T}). \tag{4.6}$$

The variable $P_{T\ell}$ is the lepton transverse momentum, $\not{E}_T$ is the magnitude of the missing transverse momentum, and $\phi_{\ell\not{E}_T}$ is the azimuthal angle between them. In the special case that $\phi_{\ell\not{E}_T} = \pi$, $\hat{\theta} = \pi/2$, we have $P_{Tl} = M/2 = \not{E}_T$ so that $M_T = M$.

There is no known SM state that contributes to high transverse mass above the *W* peak. In the absence of any signal, the lack of events can be translated into a limit on the mass of particles predicted in SM extensions containing gauge bosons that are "recurrences"

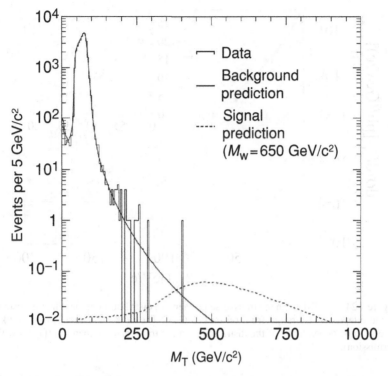

Figure 4.23 CDF data on the transverse mass of lepton plus neutrino events at high mass. The spectrum is dominated by the W boson signal at a mass $\sim$80 GeV. The predicted signal for a 650 GeV "sequential W" boson is also shown ([11] – with permission).

of the known W boson. The present data allow us to rule out sequential charged gauge bosons with a mass less than 650 GeV.

The transverse momentum of the dilepton pair in the dilepton mass range of 66 to 116 GeV, encompassing Z boson production, is shown in Fig. 4.24. It is strongly limited to low values, because it is due either to intrinsic initial state parton transverse momentum or to initial state radiation (ISR) of, say, a gluon by the quark or antiquark. In the latter case, we expect a cross section that falls as the third power of the dilepton transverse momentum. The observed distribution is at least qualitatively in agreement with that expectation.

4.7 EW production

In the previous section we looked at the continuum production of lepton pairs. Resonant production of the W gauge boson and the Z boson are prominent features of the spectrum. This large sample of singly produced gauge bosons can be used to extract some of their basic properties such as mass, decay width, and their different branching fractions. In turn, because these quantities are accurately predicted in the SM, we can test the SM to a high degree of precision.

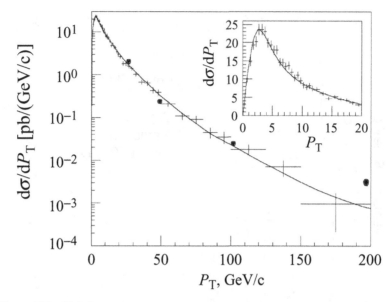

Figure 4.24   CDF data on the transverse momentum distribution of the dilepton system in the dilepton mass range from 66 to 116 GeV. The dots indicate a two-body like inverse cube dependence of the distribution on the transverse momentum ([11] – with permission).

4.7.1 W mass and width

The masses of the W and Z bosons are predicted in the electroweak theory, as discussed in Appendix A. The vacuum expectation value of the Higgs field is determined by the Fermi coupling constant, $\langle\phi\rangle = 1/\sqrt{2G\sqrt{2}} = 174$ GeV. The weak coupling constant g_W is related to the electromagnetic coupling constant e and the Weinberg angle θ_W, $G/\sqrt{2} = g_W^2/8M_W^2$, $g_W \sin\theta_W = e$. These two numbers, G, θ_W, allow us to predict the W and Z masses. In fact, these predictions were available to the experimenters prior to the data-taking runs where the W and Z were discovered at CERN in the early 1980s.

$$M_W^2 = 2\pi\alpha_W\langle\phi\rangle^2, \quad M_W \sim 80\,\text{GeV}. \tag{4.7}$$

We also saw in Appendix A that the gauge bosons have couplings to the quarks and leptons specified by the gauge principle. The coupling of the W to quarks is complicated by the existence of the weak quark mixing matrix $V_{qq'}$. However, in first approximation we can treat the mixing matrix as diagonal. Thus the W couples to all lepton–neutrino pairs, $e^- + \bar{\nu}_e$, $\mu^- + \bar{\nu}_\mu$, $\tau^- + \bar{\nu}_\tau$, and the $\bar{u} + d$, $\bar{c} + s$ quark pairs with equal (universal) strength. Note that $\bar{t} + b$ is too heavy to be a possible decay mode. We must remember to count all three possible quark colors in making a colorless final state (the W is a color singlet because color is a strong interaction attribute). These considerations lead to nine distinct dilepton or diquark final states with equal partial decay rates. The

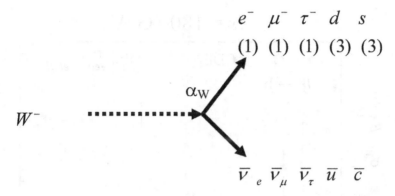

Figure 4.25 Schematic representation of the two body decays of the W boson into lepton and quark pairs. Only "diagonal" quark pairs in the quark weak mixing matrix are shown. The quark pairs each have three identical color–anticolor entries in the sum over final states, $(\overline{RR}, \overline{BB}, \overline{GG})$.

total decay rate is proportional to the weak fine structure constant α_W and the W mass.

$$\Gamma(W^- \to e^- + \bar{\nu}_e) = (\alpha_W/12)M_W \sim 0.21 \text{ GeV}.$$

$$\Gamma_W \sim 9\Gamma(W^- \to e^- + \bar{\nu}_e). \tag{4.8}$$

The total W decay width is predicted to be about 2.0 GeV. The width to mass ratio for the W boson is about 2.5 percent, which makes the W a fairly sharp resonance. A schematic representation of the W two body decays into lepton and quark pairs is shown in Fig. 4.25.

The coupling of the Z bosons to quark and lepton pairs is also specified in the electroweak theory as sketched out in Appendix A. For example, decays into neutrino–antineutrino pairs have a partial width that is also proportional to the weak fine structure constant and the Z mass. This dependence is clear from simple diagrammatic and dimensional considerations.

$$\Gamma(Z \to \nu_\ell \bar{\nu}_\ell) = [\alpha_W/24][M_Z/\cos^2\theta_W] \sim 0.16 \text{ GeV}. \tag{4.9}$$

Data from both the D0 and CDF experiments are shown in Fig. 4.26 for the production cross section, branching ratio, and decay width of gauge bosons. We expect to find a value $\sim \pi^2(\Gamma/M^3)(2J+1)(B \sim 2/9) \sim 9$ nb for the W cross section formed in $\bar{u} + d$ and $\bar{d} + u$ annihilations, where the leptonic branching ratio for electrons or muons is $B \sim 1/9 = 0.11$, and $\Gamma \sim 2.0$ GeV from Eq. (4.8). The data shown confirm these approximate expectations.

The Z mass has been measured to extremely high accuracy at electron–positron colliders located at CERN (LEP) and the Stanford Linear Accelerator Center, SLAC (SLD). Therefore, we will assume we know it to arbitrary accuracy. The W mass is more difficult to measure. As we mentioned in Chapter 1, at LEP II the production of W pairs has been measured. The shape of the cross section as a function of LEP CM energy as the "threshold" at CM energy $\sim 2M_W$ is crossed then allows for a measurement of the W

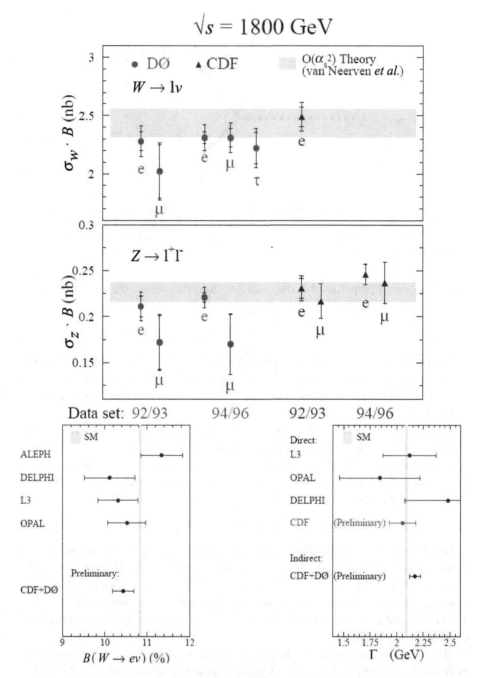

Figure 4.26   Data from the LEP, D0, and CDF experiments on the production cross section, the branching ratio, and the decay width for both W and Z gauge bosons. The vertical lines indicate the rough predictions made in the text for the total decay width and the electronic branching fraction ([12] – with permission).

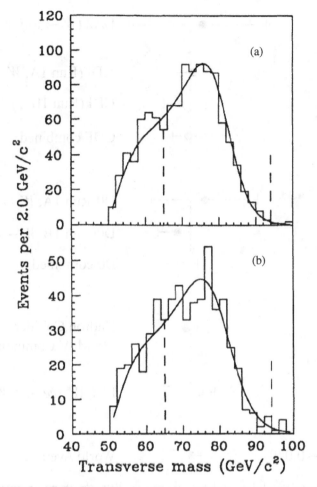

Figure 4.27 Data from the CDF experiment on the transverse mass of *W* bosons in lepton plus neutrino final states, (a) for electrons and (b) for muons. The long resonant tail at high mass is displayed in the data, which allows for a simultaneous measurement of the *W* decay width ([13] – with permission).

mass. At the Tevatron, a direct measure of the invariant mass of the *W* decay products is used to determine the mass.

By design, CDF and D0 have concentrated on the leptonic decay mode using precisely measured muons (using the tracking – Chapter 2) or electrons (precision calorimetry and/or tracking). The *Z* can be used as a control sample so that the *W*–*Z* mass difference is what is really measured and some systematic effects common to the two measurements cancel. At large transverse mass the shape is dominated by the Breit–Wigner width, since the resonant falloff with mass is much slower (power law, $[\Gamma /(M - M_0)]^2$) than the Gaussian falloff due to the error in the mass measurement. Therefore, the transverse mass distribution can be used to measure both the *W* mass and, using the high mass tail, the decay width (Fig. 4.26). In Fig. 4.27 data from the CDF experiment on the transverse mass of *W* gauge bosons are displayed.

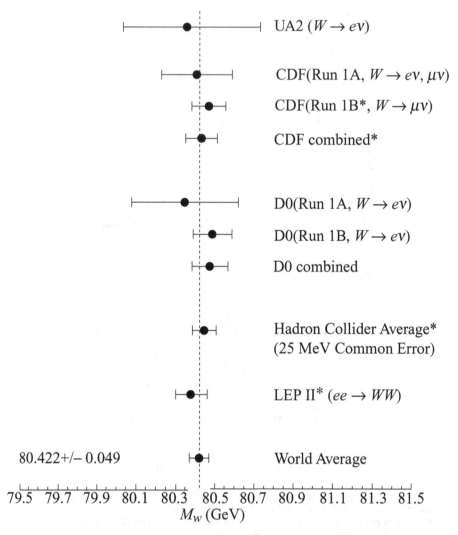

Figure 4.28 Determinations of the *W* mass from the UA2 (CERN), CDF, D0, and LEP experiments directly producing *W* gauge boson pairs ([13] – with permission).

A good knowledge of the *W* transverse momentum spectrum is also needed to measure the mass accurately, because it influences the transverse mass distribution of the *W*. It is here that the *Z*, used as a control sample, is very useful in evaluating, and thus controlling, systematic errors. The collider results on direct measurement of the *W* mass are shown in Fig. 4.28. The measurements from CDF and D0 are combined with those from *WW* production at LEP (see Chapter 1).

Finally, the current world data on the mass of the *W* are shown in Fig. 4.29. Data from proton–antiproton colliders are combined with the direct data from *WW* production in electron–positron machines as shown in Fig. 4.28. Then indirect measurements using

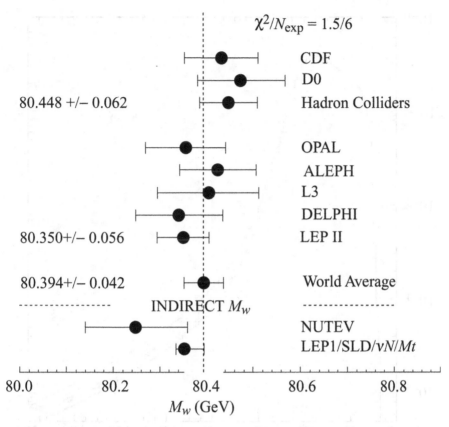

Figure 4.29 Data on the *W* mass from direct measurements at CDF, D0, and LEP and from indirect measurements using lepton scattering data ([13] – with permission).

data that depend on virtual *W* exchange or the Weinberg angle, θ_W, are combined with the direct measurements. These give the combined result. As we will see later in this chapter, precision data on the top and *W* masses can be used along with electroweak calculations of the radiative mass shift due to higher order "loop" processes to set limits on the Higgs mass.

The dependence of the distribution of transverse mass on the *W* decay width is shown in Fig. 4.30. The fractional differences arising from different decay widths are most apparent at high transverse mass, as expected. Clearly, with a sufficient number of events, an accurate measurement of the *W* width is possible (see Fig. 4.26).

4.7.2 P_T of W

The Drell–Yan production of a single *W* is a two-to-one process, with essentially no transverse momentum in the final state. As we saw with charmonium (Chapter 3) and lepton pairs (Chapter 4), this is true to lowest order, but initial state radiation will cause

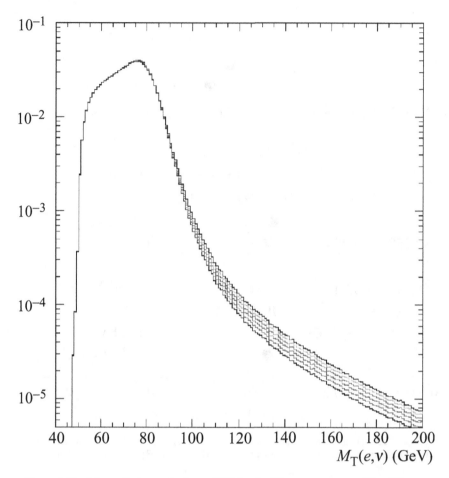

Figure 4.30 Monte Carlo results from CDF for the *W* transverse mass. The different curves correspond to *W* decay widths of 1.5 to 2.5 GeV ([14] – with permission).

the *W* to have a finite transverse momentum. As a side comment, we will see in Chapter 5 that one important mode for Higgs production arises from the radiation of a Higgs by a highly virtual *W* or *Z* gauge boson (Higgs Bremsstrahlung).

Data for single *W* production taken at the Tevatron are shown in Fig. 4.31. The transverse momentum of the *W* peaks at very low values. Although the data are for any event with a found *W*, there are very often jets found that accompany the *W*. One of the Feynman diagrams used in the COMPHEP Monte Carlo program for initial state radiation by the colored quarks is topologically just the basic two body scattering, $u + \bar{d} \rightarrow W^+ + g$. Therefore, we expect that the transverse momentum of the *W* gauge boson is distributed as the inverse cube of the transverse momentum, as we saw for the *Z* in Fig. 4.24. The line shown in Fig. 4.32 has this behavior, and we can see that it is a reasonable representation of the results of the full Monte Carlo model, at least at high transverse momenta.

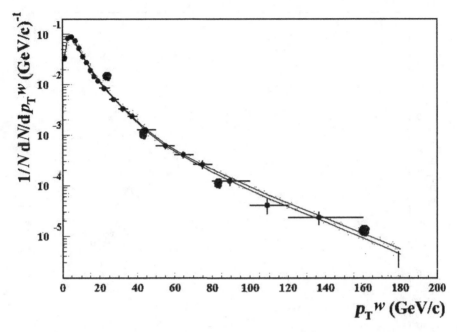

Figure 4.31 D0 data on the transverse momentum distribution of singly produced W bosons. The dots indicate a $1/P_T^3$ behavior. Initial state radiation (ISR) of a gluon by one of the initial state quarks causes a recoil transverse momentum to be taken up by the W boson. Systematic errors on the data points are also indicated ([15] – with permission).

4.7.3 W asymmetry

There is an asymmetry in the production of W bosons in proton–antiproton collisions that is due to a combination of two effects; the V-A nature of the weak interactions (see Appendix A) and the dynamics of W production. In the example of W^+ production from valence quarks shown in Fig. 4.33, the positrons are preferentially emitted in the direction of the antiproton. The similar reaction, $\bar{u} + d \rightarrow W^- \rightarrow e^- + \bar{\nu}_e$, sends electrons in the direction of the proton.

We must simply assert that the V-A, parity violating, nature of the weak interactions makes light quarks and leptons (u, d, e^-, ν_e in the first generation) left handed (negative helicity, where helicity is the projection of spin on the direction of the momentum) and the corresponding antiparticles, $\bar{u}$, $\bar{d}$, e^+, $\bar{\nu}_e$, right handed (positive helicity).

The lepton charge asymmetry can be used to study the difference in the up and down quark distribution functions of quarks in the proton. The final lepton charge asymmetry is clearly dependent both on the V-A dynamics and on the distribution of u and d quarks in the proton. Assuming that we fully understand the fundamental two body weak production and decay dynamics, we can use the data to constrain the input values for the $u(x)$ and $d(x)$ quark distribution functions.

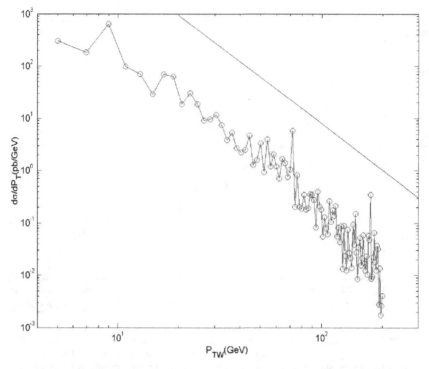

Figure 4.32   The COMPHEP distribution of *W* transverse momentum for the production of *W* gauge bosons and gluons in the final state, $u + \bar{d} \rightarrow W^+ + g$ for *p–p* interactions at the LHC. The line shows typical two-body scattering behavior, where the transverse momentum is distributed as the inverse cube, $1/P_{TW}^3$.

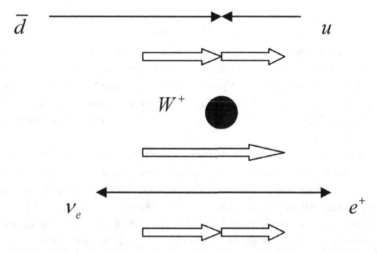

Figure 4.33   Schematic representation of the spin correlations in proton–antiproton production of single *W* gauge bosons. Momenta are indicated as thin arrows, spin directions as thick arrows. The initial state quarks are assumed to be valence. Therefore, the antiquark is along the antiproton direction and the quark is along the proton direction. Positrons are preferentially emitted in the direction of the incident antiproton.

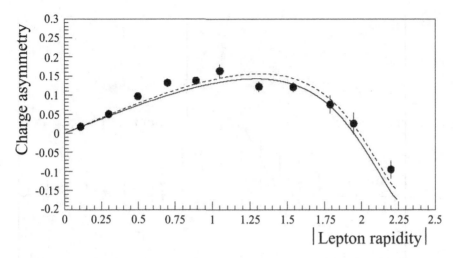

Figure 4.34   Data from CDF on the lepton charge asymmetry as a function of the lepton rapidity in the production of single *W* bosons ([16] – with permission). The curves are predictions using different quark distribution functions.

The CDF data on the lepton charge asymmetry as a function of the lepton rapidity are shown in Fig. 4.34. Subsequently, those data have been used to constrain the quark distribution functions. At large x, the value of $u(x)$ is larger than $d(x)$ even though both are valence quarks with equal binding (color). That is seemingly just an experimental fact we need to remember.

4.7.4 b pair decays of Z, jet spectroscopy

The calorimetric resolution for dijet masses is important in searches for the Higgs boson, for example for the case of Higgs decays into *b* quark pairs. Data from CDF are shown in Fig. 4.35. These data serve to indicate the mass resolution that can be obtained in jet spectroscopy. The data come from a sample of dijets with two decay vertices identified ("*b* tags", see Chapter 2). The observed mass resolution is roughly $dM \sim 12$ GeV. The error due to energy measurement can be estimated (roughly) to be 7 GeV ($a = 60\%$, see Chapter 2). Clearly, there are other contributions to the mass error that arise in defining jet energy and that lead to the total mass error. This exercise is essential practice and serves as a control sample for searches in dijet mass spectra. We will use these estimates to extrapolate to the mass resolution expected in calorimetric Higgs searches in our discussions in Chapter 5.

We can note that $\sim 20\%$ improvements are being seen in mass resolution if tracking information is used in conjunction with calorimetric measurements. This is called "energy flow" in the literature. The idea is simple. Tracking measurements of charged pion momenta are much more accurate than calorimetric energy measurements at "low" momenta (<100 GeV). Much larger improvements are expected for detectors used in electron–positron machines because there is no confusion from an underlying event.

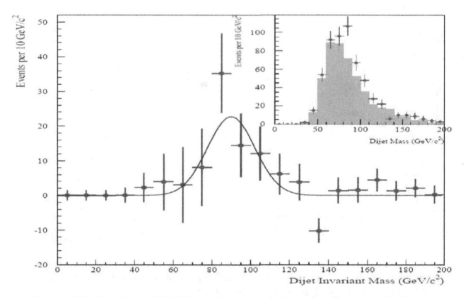

Figure 4.35   Data from CDF ([3] – with permission) on the dijet mass distribution reconstructed by calorimetry. The jets have both been tagged as *b* quark candidates using the silicon tracking detectors (see Chapter 2). The data points and the expected background are shown in the inset. The background subtracted spectrum and a Monte Carlo prediction are also shown.

2 – What is M_H and how do we measure it? (This refers to the second of the dozen questions raised in Section 1.7. We will repeat them as we get to the point of trying to address them).

4.8 Higgs mass from precision EW measurements

At this point we can finally begin to address the second unanswered question first posed at the end of Chapter 1. "What is M_H and how do we measure it?" First, however, we need to digress a bit and look at the effects of higher order quantum "loops" on observable quantities. As with charge, the operational mass of a particle (defined by the behavior of the "propagator" $\sim 1/(q^2 + M^2)$) is not a fixed number but is an effective constant in quantum field theory with a value that depends on higher order quantum processes, as discussed in Appendix D. The experimental exploration of the SM has now progressed in accuracy to the point where we can test its predictions at "one loop" in the perturbation expansion in powers of the weak coupling constant.

In Fig. 4.36 we show a schematic representation of the fermion and boson loops contributing to a propagator. Since the propagator is altered by these loops and since it has the form $V(q) = 1/(q^2 + M^2)$, to lowest order (see Eq. (1.6)), we can expect that the mass will be altered by the loop contributions. Indeed, this is correct. Conversely,

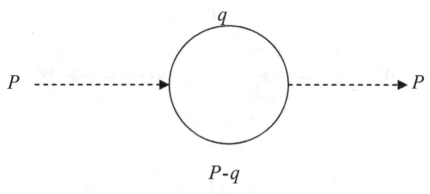

Figure 4.36 Schematic representation of the virtual decay, $p \rightarrow (p - q) + q$, and subsequent absorption, $(p - q) + q \rightarrow p$, of a pair in a "loop diagram".

measuring the mass very precisely, we learn about the particles that exist virtually in the quantum loops. In fact, we can in this way constrain the mass of the Higgs boson.

A particle propagates virtually with momentum P and then virtually decays into a pair of fermions or bosons that are reabsorbed to reform the initial particle. This is a higher order "loop" diagram. The "running" of the coupling "constants" is also due to higher order quantum loop corrections and is discussed in Section 4.2 and Appendix D.

We must simply assert that the propagators for fermions (Dirac Equation) and bosons (Klein–Gordon Equation) are different, $1/q$, $1/q^2$ respectively, for massless quanta. We have already mentioned, Eq. (1.6), that the propagator for massless bosons can be thought of as the Fourier transform of the Coulomb interaction potential. The propagator for fermions follows from a study of the massless Dirac Equation (see the references given at the end of Chapter 1 and Appendix A).

The expressions for the modification to the propagator (or mass squared) of particle P due to fermions and bosons in the loop come after integrating over all possible virtual loop momenta. We see that for fermions the integral goes as the square of the fermion mass, m, while for bosons it has a much weaker dependence, going as the logarithm of the boson mass, M.

$$\delta M^2 \sim \int d^4q/(q)^2 \sim \int q^3 dq/q^2 \sim \int^m q\,dq \sim m^2,$$

$$\delta M^2 \sim \int d^4q/(q^2)^2 \sim \int q^3 dq/q^4 \sim \int^M dq/q \sim \ln(M). \qquad (4.10)$$

The Higgs mass is a free parameter in the current "Standard Model" (SM). There are two parameters in the Higgs potential, and one is fixed by the measurement of the vacuum expectation value of the Higgs field using the Fermi coupling constant, G. The other can be taken to be the Higgs mass, and it too is not determined by theory and must

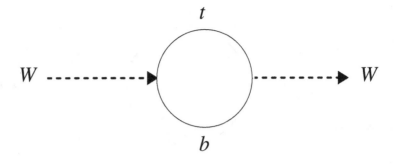

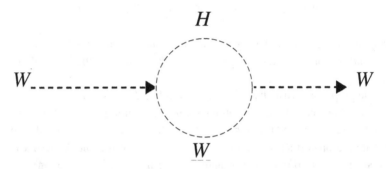

Figure 4.37   Loop diagrams for the virtual W decays that contribute to the W boson mass. There are both quarks, b and t, and gauge bosons, W and H, in the intermediate states. The couplings are Wtb and WWH.

be determined from experiment. Precision data taken on the Z resonance do, however, constrain the Higgs mass. The Z mass is known very well. The top and W masses are determined as we discussed in this chapter, $m_t = 176 \pm 6 \, \mathrm{GeV}$, $M_W = 80.41 \pm 0.09 \, \mathrm{GeV}$. Both measurements are statistics limited at present, so we can expect improvements in the near future as CDF and D0 gather more data.

The SM at lowest order predicts that $M_Z = M_W / \cos \theta_W$, as we show in Appendix A. Radiative corrections due to loops will modify this relationship because for Z loops there are top pairs of fermions, while for W^+ loops the pair is $t + \bar{b}$. Therefore the mass in the loop differs, causing a differential shift of the Z mass with respect to the W mass.

A schematic representation of the important loop diagrams for W gauge bosons is shown in Fig. 4.37. The best determined parameters in the SM for the electroweak interaction are G (muon decay), the Z mass (LEP), the fine structure constant, α, and the Weinberg angle, θ_W (neutral current neutrino interactions, Z lepton and quark decay asymmetries). These parameters are sufficient to predict the W mass up to radiative corrections due to top loops and Higgs loops. The program is then to precisely measure the W mass and the top mass and thus constrain the Higgs mass.

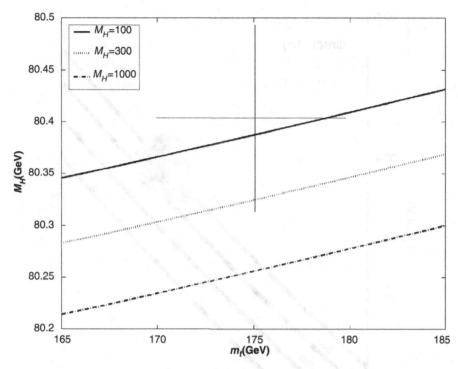

Figure 4.38 Data from the Tevatron experiments using both the direct top quark mass measurements and the precision W mass measurements to constrain the Higgs mass.

The expression for the shift of the squared W mass due to the fermion and boson loops is given in Eq. (4.11). We see the expected quadratic mass dependence for the fermions and the logarithmic mass dependence for the bosons in the loop. There are opposite signs for the contributions to mass from fermion and boson loops. This sign difference will be crucial in our discussion of SUSY in Chapter 6. As the top mass increases the W mass increases (fermions), while the W mass decreases as the Higgs mass increases (bosons).

$$M_W^2 = M_Z^2 \cos^2 \theta_W (1 + \delta),$$
$$\delta_t \sim [3\alpha_W(m_t/M_W)^2]/16\pi,$$
$$\delta_H = -[11\alpha_W \tan^2 \theta_W/24\pi]\ln(M_H/M_W). \qquad (4.11)$$

The explicit sensitivity of the W boson mass to the top mass is:

$$dM_W = (3\alpha_W/16\pi)(m_t/M_W)dm_t. \qquad (4.12)$$

For example, the present top uncertainty of $\sim$5 GeV in mass leads to a 22 MeV shift in the W mass. The dependence on the Higgs mass is much weaker. For a Higgs mass between 100 and 1000 GeV, the W mass shifts by only 130 MeV. The student is strongly encouraged to put some numbers into Eq. (4.11) in order to get a feel for the sensitivity involved. The result of plugging in the numbers is shown in Fig. 4.38. Clearly,

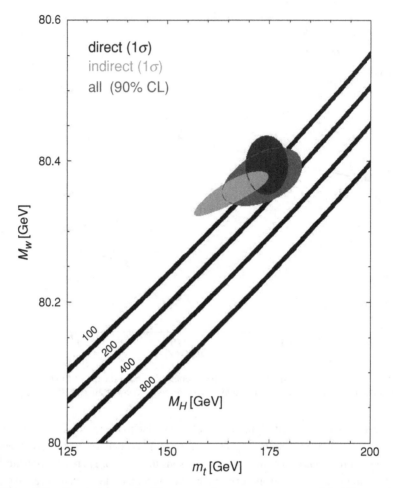

Figure 4.39   Constraints on the Higgs mass due to measurements of the *W* mass, the
top mass, and the other precision electroweak data ([17] – with permission).

an accuracy of 25 MeV on the *W* mass (∼0.3%) or better is needed to restrict the Higgs
mass for a light Higgs to 100 GeV in the context of the SM.

A more comprehensive compilation of all the presently available precision data is
shown in Fig. 4.39. The direct measurements of the top and the *W* masses are plotted.
Indirect measurements of electroweak parameters are shown as a separate allowed region.
These two sets of independent measurements are not particularly consistent. We might
worry that combining data and thereby reducing the errors is perhaps not a good idea
because of systematic uncertainties.

In addition, the contours plotted only include one standard deviation (68% confidence
level) instead of the more conventional two standard deviation contours. In any case, it
appears that a light Higgs mass is "favored" by the existing electroweak data if the SM
is a correct theory. Recall that LEP2 data set a limit that $M_H > 113$ GeV. Clearly, more

data with higher statistics, which will eventually be available from CDF and D0, will tell us whether the prediction of a low mass Higgs boson persists and is made sharper or if the data becomes accurate enough to indicate an inconsistency with the SM, such as $M_W = 80.55$ GeV, $M_H > 115$ GeV.

Note, however, that this analysis assumes that the Standard Model is a fundamental theory, while it is felt by many, because of the unanswered questions posed in Chapter 1, to be incomplete and thus only an effective field theory. Therefore, the derived constraints on the Higgs mass are not logically self consistent. A more general analysis makes for much less restrictive Higgs mass constraints. We must be careful to avoid making glib arguments when looking at unknown phenomena. Clearly, a strong statement about the Higgs mass is not possible at present.

Exercises

1. Use the formulae developed in Chapter 3 to estimate the cross section in p–p collisions at 2 TeV CM energy for g–g scattering at a mass of 200 GeV. Compare the result with the data shown in Fig. 4.3. (Use $\Delta y = 4$, C $= 1$).
2. Show that $d\chi/d\hat{t} \sim 1/\hat{t}^2$.
3. Plot $\alpha_s(Q^2)$, $\Lambda_{QCD} = 0.2$ GeV from 1 GeV to 1 TeV. Compare with Fig. 4.7.
4. Look at the plot of the strong coupling constant in COMPHEP. Compare it with Fig. 4.7.
5. Use Table 3.1 to estimate b quark pair production with respect to jet production.
6. Evaluate the muon lifetime in Eq. (4.4) and compare with the experimental value of ~ 2.2 μs. (NB $\hbar = 6.6 \times 10^{-25}$ GeV s).
7. Find the muon decay width in COMPHEP and compare with the result of Exercise 6.
8. Evaluate the top decay width using Eq. (4.5).
9. Use COMPHEP to evaluate the top decay width, and compare with Exercise 8.
10. Make the numerical calculation shown in Eq. (4.7) for the W decay width.
11. Evaluate the loop contribution of the Higgs to the W mass, Eq. (4.10), for Higgs masses of 100, 300, and 1000 GeV and compare with Fig. 4.36.
12. Differentiate the expression for the W mass to show that $dM_W/M_W = [-11\alpha_W \tan^2 \theta_W / 48\pi](dM_H / M_H)$.
13. Evaluate the expression derived in Exercise 12 to show that $dM_W \sim 57$ MeV (dM_H / M_H). For Higgs mass from 100 to 1000 GeV take the fractional Higgs mass variation to be ~ 3 with respect to the mean of ~ 300 GeV and compare with Fig. 4.36.
14. Use COMPHEP to evaluate $g + g \rightarrow t + \bar{t}$ in proton–antiproton collisions at 900 GeV + 900 GeV CM energy. Compare with the data given in this chapter. Try $u + \bar{u}$ quark annihilation into top pairs. Is the cross section larger? Why?
15. Use COMPHEP to evaluate the radiative width of the W, $W^+ \rightarrow E1$, $n1$ and $W^+ \rightarrow E1$, $n1$, A. What fraction of two body W decays have a photon emitted by the electron?
16. Evaluate the Drell–Yan process, u, $U \rightarrow e1$, $E1$ for proton–antiproton collisions at 2 TeV CM energy for masses >50 GeV. Compare with the data given in this chapter.

References

1. Montgomery, H., Fermilab-Conf-99/056-E (1999).
2. Blazey, G. and Flaugher, B., Fermilab-Pub-99/038 (1999).
3. Montgomery, H., Fermilab-Conf-98-398 (1998).
4. Particle Data Group, *Review of Particle Properties* (2003).
5. Bethke, S., MPI-PhE/2000-02 (2000).
6. Huth, J. and Mangano, M., *Ann. Rev. Nucl. Part. Sci.*, **43**, 585 (1993).
7. D0 Collaboration, *Phys. Rev. Lett.*, **74**, 3548 (1995).
8. Quigg, C., Fermilab-FN-676 (1999).
9. Tollefson, K. and Varnes, E., Direct measurement of the top quark mass, *Ann. Rev. Nucl. Part. Sci.*, **49**, (1999).
10. CDF Collaboration, *Phys. Rev. Lett.*, **82**, 4773 (1999).
11. Seidel, S., Fermilab-Conf-01/054-E (2001).
12. Steinbruck, G., Fermilab-CONF-00/094-E (2000).
13. Lancaster, M., Fermilab-Conf-99/366-E (2000).
14. Baur, U., Ellis, R. K., and Zeppenfeld, D., Fermilab-Pub-00/297 (2000).
15. D0 Collaboration, *Phys. Rev. Lett.*, **80**, 5498 (1998).
16. Glenzinski, D. and Heintz, U., arXiv:hep-ex/0007033 (2001).
17. Pitts, K., Fermilab-Conf-00-347-E (2001).

Further reading

Altarelli, G. and Di Leuo, L., *Proton–Antiproton Collider Physics*, Directions in High Energy Physics, Vol. 4, Singapore, World Scientific (1989).

The collider detector at Fermilab: collected physics papers, Fermilab Pub. 90/31-E (1990).

The collider detector at Fermilab: collected physics papers, Fermilab Pub. 91/60-E (1991).

The collider detector at Fermilab: collected physics papers, Fermilab Pub. 92/ 138-E (1992).

The D0 Experiment at Fermilab: collected physics papers, Fermilab Pub-96/064-E (1996).

Ellis, R. K. and W. J. Sterling, *QCD and Collider Physics*, Fermilab Conf. 90/167–7 (1990).

Hinchliffe, I. and A. Manohar, The QCD Coupling Constant, *Ann. Rev. Nucl. Part. Sci.*, **50**, 643 (2000).

Pondrom, L., *Hadron Collider Physics*, Fermilab Conf. 91/275-E (1991).

Schochet, M., The Physics of Proton–Antiproton Collisions, Fermilab Conf. 91/341-E (1991).

Shapiro, M. and J. Siegrist, Hadron collider physics, *Ann. Rev. Nucl. Part. Sci.*, **41**, 97 (1991).

Thurman-Keup, R. M., A. V. Kotwal, M. Tecchio, and A. Byon-Wagner, W boson physics at hadron colliders, *Rev. Mod. Phys.*, **73**, April (2001).

UA1 Collaboration, G. Arnison *et al.*, *Phys. Lett. B* **122**, 103 (1983).

UA2 Collaboration, *Phys. Lett. B* **241**, 150 (1990).

5

Higgs search strategy

You may seek it with thimbles – and seek it with care; . . . you may charm it.
The Hunting of the Snark – Lewis Carroll

Come Watson, the game is afoot.
Sherlock Holmes – Sir Arthur Conan Doyle

We are now ready to examine the experimental search strategies for first discovering the Higgs boson and then finding out whether its properties are what we expect if the SM is correct. For example, is the coupling to W and Z bosons as predicted? Does the coupling to fermions and leptons go as the fermion mass? Are the self-couplings of the Higgs as predicted? New experiments being prepared for the LHC at CERN are explicitly designed to attempt to answer as many of those questions as possible.

The expected properties of the Higgs boson were first mentioned in Chapter 1 and Appendix A. The accuracy of the measurements of the SM particles into which the Higgs decays was explored in Chapter 2. The formulae needed to calculate p–(anti)p production cross sections were developed in Chapter 3 and the hadron collider state of the art was presented in Chapter 4. We now put all of this information together in order to look at the production and decay of the Higgs boson, the last undiscovered particle in the SM "periodic table." We want to find the mass, width, quantum numbers, couplings to fermions and gauge bosons, and self-couplings of the Higgs boson.

5.1 Cross sections at the LHC

We first mention the "minimum bias" rates for "inclusive" or unselected inelastic events at the LHC. The expected total inelastic cross section is $\sigma_1 \sim 100$ mb, of which ~ 50 mb is not "diffractive" in character. Diffractive events send a scattered proton at small angles to one or the other or both of the incident proton beams. We assume here that these scattered protons exit at angles less than those covered by our detectors. There are specialized experiments that will run at the LHC which will detect these low transverse momentum protons in order to study the elastic and diffractive interactions. In what follows we specialize to non-diffractive high transverse momentum reactions.

In Chapter 3 we first mentioned two-to-one resonance production. In the narrow width approximation, these processes have fundamental cross sections, as shown in Eq. (5.1).

$$\hat{\sigma} \sim \pi^2(2J + 1)\Gamma/M^3. \tag{5.1}$$

For example, the Drell–Yan production of W bosons can be estimated, with $M = M_W$, $J = 1$, and $\Gamma = \Gamma_W$, to be $\hat{\sigma}_W \sim 47\,\mathrm{nb}$.

For any fundamental two body scattering process a rough approximation for the cross section for production of pair of particles of mass M_0 is:

$$\Delta\hat{\sigma} \sim \pi\alpha_1\alpha_2/(2M_0)^2. \tag{5.2}$$

For two body scattering, the point like scattering dynamics leads to a mass distribution, $d\hat{\sigma}/dM$, which goes as the inverse cube of the mass. Integrating that distribution above a threshold at $2M_0$, we are led to Eq. (5.2). For example, WW production is estimated to be:

$$\Delta\hat{\sigma}_{WW} \sim \pi\alpha_W^2/(2M_W)^2 = 50\,\mathrm{pb}.$$

In comparison to the inelastic non-diffractive cross section, the Higgs production cross section is very small ($\sigma_I \sim 50\,\mathrm{mb}$, σ_H (120 GeV) $\sim 20\,\mathrm{pb}$), in the ratio of 4×10^{-10}. Because the Higgs cross section is so small, we must have high luminosity and that, in turn, means an enormous rate of particles from uninteresting "minimum bias" or inelastic, non-diffractive events must be sorted through.

The last quark discovered was the top quark, found at the Tevatron. The CDF and D0 experiments successfully found the top quark, which has a cross section $\sim 10^{-10}$ of the total cross section, so finding such rare processes has a precedent.

The cross section for various processes in p–(anti)p collisions is shown as a function of CM energy in Fig. 5.1. The cross section for Higgs masses other than 500 GeV can be extracted from Fig. 5.3. For the LHC we will assume a design luminosity of $10^{34}/\mathrm{cm}^2$ s. For one year of running we put in an efficiency of $\sim 1/3$ or a data taking time of 10^7 s. This means a sensitivity of $10^{41}/\mathrm{cm}^2$ yr or 100/fb^{-1}yr. In one year at design luminosity 100 000 (1 000 000) Higgs particles of 500 GeV (100 GeV) mass will be produced. Note that the cross section for top at the Tevatron is about the same magnitude as a 100 GeV Higgs at the LHC. Nevertheless, because the Higgs mass is unknown and could be up to about 1 TeV, the LHC accelerator and detectors must prepare to explore cross sections much lower than those probed at the Tevatron.

A 500 GeV mass Higgs has a production cross section 1000 times smaller at the Tevatron than at the LHC, but the inelastic cross section is roughly the same at the two energies. Even at the LHC, a 500 (100) GeV Higgs has a fractional cross section with respect to the inelastic cross section of only $\sim 10^{-11}(10^{-10})$, which requires great rejection power against backgrounds and a high luminosity. This rejection must exceed what has presently been achieved for the top quark. As we noted in Chapter 2, multiple redundant measurements of the SM particles will be required if the needed rejection power is to be achieved in the LHC experiments.

We also see that the cross section for the strong production of top pairs, each top decaying into $W + b$, rises very rapidly from the Tevatron to the LHC, as we mentioned in Chapter 4. These top pairs will make a background of W pairs, which will complicate our Higgs searches when we are trying to measure the Higgs decaying into WW. The

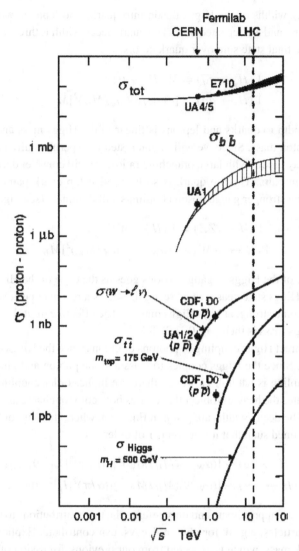

Figure 5.1   Cross section for all, *b* quark, *W*, *t* quark, and 500 GeV Higgs particles as a function of the CM energy ([1] – with permission).

top cross section exceeds that for a 500 (100) GeV Higgs boson by a factor ∼300 (10), so excellent top rejection is needed.

5.2 Higgs direct and "loop" couplings

In Appendix A we derive the coupling of the Higgs field to the gauge bosons. We also postulate the Yukawa coupling of the Higgs to the fermions, and find that the coupling constants are proportional to the masses of the fermions. These couplings then imply calculable decay widths of the Higgs boson into quarks and leptons, which were first given in Chapter 1 and are repeated here. The quark decay width is three times the lepton width due to the final state sum over quark colors.

$$\Gamma(H \to q\bar{q}) = 3\Gamma(H \to l\bar{l}),$$
$$\Gamma(H \to q\bar{q}) = [(3\alpha_W/8)(m_q/M_W)^2]M_H. \tag{5.3}$$

The decay width to quarks and leptons is linear in the Higgs mass and quadratic in the quark or lepton mass. Since we will be interested in experimentally tractable Higgs decays into decay modes with large branching ratios, we will consider decays to b quark pairs or τ lepton pairs. The top quark is so heavy that top quark pairs are above ZZ threshold, but the stronger gauge boson couplings still dominate (see Fig. 5.15).

$$\Gamma(H \to ZZ) = \Gamma(H \to WW)/2,$$
$$\Gamma(H \to WW) = [(\alpha_W/16)(M_H/M_W)^2]M_H. \tag{5.4}$$

The coupling of the Higgs to gauge bosons goes as the cube of the Higgs mass. This means that the Higgs state ceases to be recognizable as a resonant peak when the weak interactions become strong, at high Higgs masses. The effective limit, $\Gamma_H \sim M_H$, on the observable Higgs mass is then $\sim(1.0-2.0)$ TeV.

There is no direct Higgs coupling to photons or gluons since the Higgs has no electric charge or color. Since the Higgs couples to mass and the photon and gluons are massless, that decoupling is natural. However, there are higher order couplings. We use as intermediate states the heaviest object that carries both color or charge and weak charge, the top quark. The decay widths are given in Eq. (5.5), where the symbol $|I|$ indicates a loop integral defined such that it is a number of order one.

$$\Gamma(H \to gg) \sim [(\alpha_W/8)(M_H/M_W)^2][(\alpha_s/\pi)^2|I_g|^2/9]M_H,$$
$$\Gamma(H \to \gamma\gamma) \sim [(\alpha_W/9)(M_H/M_W)^2][(\alpha/\pi)^2|I_\gamma|^2/9]M_H. \tag{5.5}$$

These results are approximate and only refer to the top contributions to the loop while several other particles, e.g. W for photon decays, can contribute. Explicit dependence on the top quark mass, which is expected from our previous discussion of fermion loop contributions to the W and Z mass, is contained in the loop integrals and is not shown here.

These loop decay widths look like allowed decays, Eq. (5.4), but with an additional factor given in the right-most brackets containing the loop integral $|I|$ and the strong or electromagnetic fine structure constant squared. That latter factor comes from the two added vertices shown schematically in Fig. 5.2. Clearly, we can think of these higher order decay modes as being due to a Higgs decaying virtually into a top pair followed by quark radiation of two photons (two gluons), leading to the $\alpha^2(\alpha_s^2)$ factor in the decay width.

Numerically, for a Higgs boson of 150 GeV mass, the gluon–gluon decay width is ~ 0.25 MeV and the two photon width is ~ 1.16 keV if we ignore the loop integral, $|I|$.

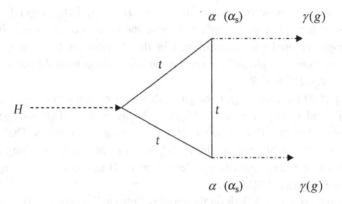

Figure 5.2   Schematic representation of the top loop decay of the Higgs boson into two photons (two gluons).

In comparison the b pair direct decay width, Eq. (5.3), using 4.5 GeV for the b quark mass, is ~6 MeV.

The effective Higgs gluon coupling constant is $\alpha_{gg} \sim (\alpha_W/9)(\alpha_s/\pi)^2$. Note that COMPHEP does not contain loop diagrams, so that these indirect decay modes are not present in COMPHEP. However, an effective ggH or $\gamma\gamma H$ interaction may be added to the Standard Model vertices by editing the COMPHEP file. The interested student is encouraged to attempt this feat.

We saw in Chapter 3 that the proton consists of u quarks, d quarks, gluons, and a "sea" of quark pairs at low x values. The masses of the u and d quarks are both ~MeV (see Fig. 1.2). Therefore, given the quadratic dependence of the Higgs width on quark masses, Eq. (5.3), the coupling of the Higgs to ordinary matter is very weak. Likewise, the coupling to the massless gluons is higher order in the coupling constants and correspondingly weak. Thus, it is hard to produce a Higgs boson in p–p collisions. The major production mechanism at the LHC is the higher order process of two gluons forming a Higgs via an intermediate top quark loop because the gluons are copiously available in the proton at low x. Thus the most important production mechanism involves particles (gluons) that do not even couple to the Higgs at lowest order in the coupling constants.

5.3 Higgs production rates

5.3.1 gg fusion

In Chapter 3 we derived formulae for the cross section for two-to-one processes. We recall the kinematics, $x_1 x_2 = M_H^2/s$, and for production at rest in the CM system, $x_1 = x_2 = \langle x \rangle = M_H/\sqrt{s}$. For a light Higgs at the LHC with CM energy = 14 TeV, the x values are small. For example, a 150 GeV Higgs is produced by gluons with proton momentum fraction $\langle x \rangle \sim 0.011$.

The formation cross section is $d\sigma/dy \sim \pi^2\Gamma(H \rightarrow gg)/(8M_H^3)$ $[xg(x)]_{x1}$ $[xg(x)]_{x2}$. The 1/8 color factor has been applied because the produced Higgs is colorless and there are eight colored gluons subsumed in the distribution function $g(x)$. Of the 8×8 combinations of a gluon from one proton and a gluon from the other, only eight are colorless, e.g. $R\overline{G} \otimes G\overline{R}$.

Using Eq. (5.5) for $\Gamma(H \rightarrow gg)$, the gluon distribution mentioned already, $[xg(x)] = (7/2)(1-x)^6$, and taking $x_1 = x_2 = M_H/\sqrt{s}$, $d\sigma/dy \sim 49\pi^2[\Gamma(H \rightarrow gg)/(32M_H^3)]$ $[(1 - M_H/\sqrt{s})^{12}] \sim 49\pi^2\Gamma(H \rightarrow gg)/(32M_H^3)$, assuming a light Higgs. The M_H^3 behavior of $\Gamma(H \rightarrow gg)$ roughly cancels the $1/M_H^3$ behavior of $d\sigma/dy$, resulting in a Higgs cross section that is approximately independent of Higgs mass. For a light Higgs, $d\sigma/dy \sim 49|I|^2\alpha_s^2\alpha_W/[2304M_W^2]$.

Numerically, $d\sigma/dy \sim 443$ fb on the rapidity "plateau" where $y \sim 0$, if $|I| \sim 1$, or $\sigma \sim 2.2$ pb ($\Delta y \sim 5$) for a light Higgs at CMS. This agrees very roughly with the complete results shown in Fig. 5.3. Note that we do not expect good agreement because the loop integral $|I|$ has some residual dependence on the Higgs mass. For a design luminosity of $10^{34}/\text{cm}^2$ s or $\sim 100/\text{fb}$ yr, CMS will produce $\sim 200\,000$ light Higgs/yr. The high luminosity is required for a statistically convincing discovery once the effects of detection efficiency and decay branching fraction to a particular final state are taken into account.

Suppose we look at the experimentally clean signature, $H \rightarrow ZZ \rightarrow$ four leptons. There will be two narrow dilepton mass peaks at the Z mass. The experimental resolution for the Higgs mass is also quite good, since accurate tracking measurements of the lepton momenta are available. Using Eq. (5.4), the branching fraction into Z pairs is $\sim 1/3$. Since the branching fraction of Z into electron or muon pairs is 7% (the student can verify this using COMPHEP, $Z \rightarrow 2*x$), if we assume a fully efficient triggering, detection, and reconstruction efficiency, we find that in one year of data taking at design luminosity, the number of signal events, S, is 327 Higgs decays into the four lepton final state. If there were no background, $B = 0$, the signal would be an $18 = \sqrt{327} = \sqrt{S}$, standard deviation effect, which is a "convincing" discovery. The probability of finding a result within one standard deviation, for Gaussian errors, is 68%. A two-sigma effect is 95%, and a three-sigma effect is 99.7% probable. Most physicists "believe" an effect of five standard deviations or larger if the systematic errors appear to be under control (see Fig. 5.36) based on long experience with statistical fluctuations.

We show complete Monte Carlo results for the Higgs production cross section as a function of Higgs boson mass in Fig. 5.3. The dominant mechanism is gluon–gluon fusion as expected. In our approximate order of magnitude estimates given above for a light Higgs we have ignored the dependence of $|I|$ on the Higgs mass (n.b. there is a local peak in the cross section shown in Fig. 5.3 at about twice the top mass where the loop integral becomes a maximum). We also ignored the additional contributions of particles other than top in the Hgg loop. Finally, we ignored the falloff of the gluon distribution functions at larger x and the rise of $[xg(x)]$ at low x, which was mentioned in Chapter 3. All these effects contribute to the mass dependence exhibited in Fig. 5.3.

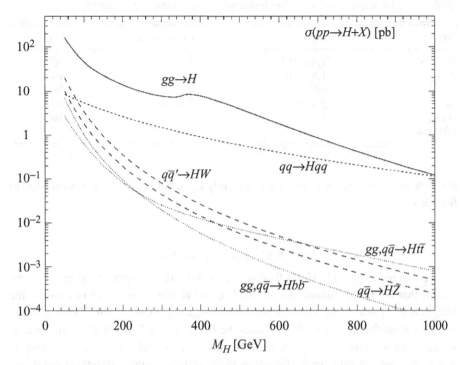

Figure 5.3   Cross section for the production at the LHC of a Higgs boson as a function of its mass. The main production process is *gg* fusion, but rarer processes are also indicated ([2] – with permission).

We will adopt representative masses for the Higgs of 120, 150, 300 and 600 GeV, with *g–g* fusion cross sections of ∼30, 20, 10, and 2 pb respectively in what follows. For these masses we estimate the total Higgs width to be 3/2 times the *WW* decay width, or ∼0.0, 1.6, 13.2, and 105 GeV. At ∼120 GeV the total Higgs decay width is very small because the Higgs mass is below the *WW* threshold ∼$2M_W$ ∼ 160 GeV. It should be clear that the final state Higgs decay mode to be used and the expected rates are very dependent on the Higgs mass. Because this mass is unknown over a rather wide mass range, we must create a flexible search strategy in order to be moderately sure of being successful.

There are 4 000 000 to 200 000 Higgs events produced per year for masses from 120 to 600 GeV according to Fig. 5.3. Using the experimentally clean *ZZ* decay mode, for Higgs bosons above *ZZ* threshold, with masses from 180 to 600 GeV, there are, for $H \to ZZ \to 4\ell$ decays, ∼8000 to 800 four lepton Higgs signal events produced per year at full LHC luminosity. This leads to a resonant signal that is detectable with a high level of statistical significance, as we will see below. In the extreme case of no background, $\sqrt{S}$ ranges from 89 to 28 standard deviations.

Even at one tenth of design luminosity, the LHC provides several discovery possibilities, as shown in Table 5.1. For example, the enormous numbers of produced *b* quarks makes the LHC a true "*b* factory." Higgs particles are discoverable in a single year even

Table 5.1 *LHC event rates for "low luminosity" operation at* $L = 10^{33}/cm^2$ *sec.*

Process	σ(pb)	Events/second	Events/year
$W \to e\nu$	1.5×10^4	15	10^8
$Z \to e^+ e^-$	1.5×10^3	1.5	10^7
$t\bar{t}$	800	0.8	10^7
$b\bar{b}$	5×10^8	5×10^5	10^{12}
H ($m_H =$ 700 GeV)	1	10^{-3}	10^4

at this reduced luminosity if they are sufficiently light, but not too light, roughly 130 to 600 GeV.

5.3.2 WW fusion and "tag" jets

Before looking at possible Higgs final states we will briefly explore production mechanisms that are not dominant. We do this because ultimately we want to measure the Higgs coupling to as many quarks, leptons, and gauge bosons as possible. The g–g fusion production mechanism basically measures the *Htt* coupling. That coupling will be convoluted with whatever couplings lead to the final state we study. The g–g mechanism is also sometimes not sufficiently distinctive to allow us to cleanly extract a Higgs decay signal into a particular final state because of large backgrounds.

In that case, we can use other, more distinctive, production mechanisms, which are biased toward rarer electroweak production processes. Additional rejection power against background can sometimes be obtained by using the characteristics of Higgs bosons; preferential coupling to gauge bosons and to high mass quarks and leptons. We will see that, for example, use of the *WW* fusion process with detected "tag jets" allows us access to Higgs decays into *W* pairs and τ pairs, which are buried in large backgrounds if only gg fusion production is considered.

Thus, by using different production mechanisms, other Higgs decay modes can be measured in addition to the "clean" rates to *Z* pairs and thence into four charged leptons or the rate into photon pairs. These two are the only decay modes available using the dominant Higgs production process, gg fusion. Obviously, improvements are of crucial importance because we aim not just to discover the Higgs boson, but also to measure as many of its properties as we can in an attempt to observe new physics beyond the SM.

The "*WW* fusion" mechanism refers to the virtual emission of a *W* boson by a quark, e.g. $u \to W^+ + d$, from both incident protons, followed by the inverse decay, or fusion, of the Higgs to a *W* pair. This mechanism is illustrated in Fig. 5.4. Clearly, this is a useful process to measure in its own right as it depends on the *HWW* coupling, compared to the gluon fusion which depends on the *Htt* coupling. The recoil jets are emitted at small angles to the proton direction and are called "tag" jets because they are an indication, or tag, that a virtual *W* was emitted.

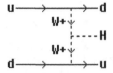

Figure 5.4   COMPHEP Feynman diagram for the production of the Higgs boson in association with recoil "tag" jets from virtual W emission.

The "WW fusion" mechanism is very similar to the analogous process where electrons or positrons emit photons. The final state in this case is any state that can be formed from two photons. The process is "tagged" by the existence of two recoil electrons in the final state emitted at small angles with respect to the incident beam. The produced state has the quantum numbers of two photons, $C = 1$ and $J^{PC} \sim 0^{++}, 2^{++}$. By the same reasoning, if a Higgs weak decay mode into two photons is established, then we will know that the Higgs spin cannot be one (Yang's Theorem).

Some LEP data for two photon production are shown in Fig. 5.5. The resonant states that are produced are "filtered" by the production mechanism to have only the quantum numbers available to diphotons.

In Chapter 3 we argued that radiation is soft and collinear. Thus, we expect that the tag jets in WW fusion have low transverse momentum, about one half the W mass, and large longitudinal momentum. The pseudorapidity distribution of the tag jets was already shown in Chapter 2 in the discussion of the required angular coverage for a typical detector operating at the LHC.

The distribution function, $f_{q/W}(x)$, for W emission by a quark q is calculable in perturbation theory, $[xf_{q/W}(x)] \sim (\alpha_W/4\pi)$, where the basic radiative behavior, $[xf(x)] \sim$ constant, is evident. For a WW mass of M and a quark pair "parent" mass of $\sqrt{\hat{s}}$, the kinematics is familiar from the similar situation worked out in Appendix C, $\tau = M^2/\hat{s}$, $x_1x_2 = \tau$. The integral representing the joint probability of emitting a W from one proton and another W from the other proton at a WW mass M is given as I_{WW} in Eq. (5.6).

$$I_{WW} = \int_{\tau}^{1} f_{q/W}(x_1) f_{q/W}(\tau/x_1) \mathrm{d}x_1/x_1$$
$$\sim (\alpha_W/4\pi)^2 (1/\tau) \ln(1/\tau). \tag{5.6}$$

The resulting fundamental cross section, using the WW Higgs decay width given in Eq. (5.4), is similar in form to the estimate we made for the gluon–gluon formation of Higgs bosons.

$$\hat{\sigma}(q\bar{q} \to q'\bar{q}'\,WW \to q'\bar{q}'\,H) \sim 16\pi^2 (\Gamma\tau/M^3) I_{WW}$$
$$\sim [(\alpha_W)^3 \ln(1/\tau)]/16M_W^2. \tag{5.7}$$

In electron–positron collisions the two photon cross section, $e^+ + e^- \to e^+ + X + e^-$, exceeds the one photon cross section, e.g. $e^+ + e^- \to q + \bar{q}$, for CM energy in the few GeV range and above. The ratio of the cross sections for the two analogous processes in p–p collisions is proportional to the ratio of the strong to weak fine structure constant squared,

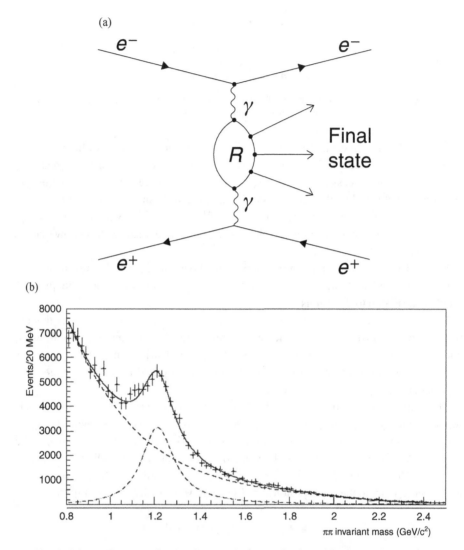

Figure 5.5   (a) Feynman diagram for two photon production of final states in electron–positron collisions. (b) The mass spectrum of the two pion final state is also shown indicating significant resonant production ([4] – with permission).

times factors of order one. Since the ratio, $(\alpha_W/\alpha_s)^2$, is only $\sim 1/9$, the *WW* fusion process is expected to be a substantial fraction of the full Higgs production cross section.

$$\hat{\sigma}(gg \to H) \sim \pi^2 \Gamma(H \to gg)/M^3 = [\alpha_W \alpha_s^2 |I|^2]/\left(72 M_W^2\right),$$
$$\hat{\sigma}(gg \to H)/\hat{\sigma}(q\bar{q} \to q\bar{q}H) \sim [(\alpha_s/\alpha_W)^2 |I|^2]/4\ln(1/\tau). \quad (5.8)$$

Indeed, as seen in Fig. 5.3, the *WW* fusion cross section is always more than $\sim 10\%$ of the gluon–gluon fusion cross section. Therefore, experimental search strategies using the *WW* fusion process are useful at the LHC.

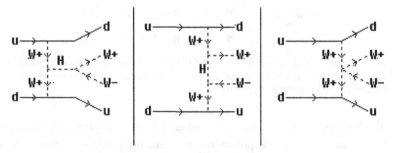

Figure 5.6   COMPHEP diagrams for *WW* fusion production of *W* pairs. Note that there are irreducible background processes. In particular, note the quartic *W* coupling.

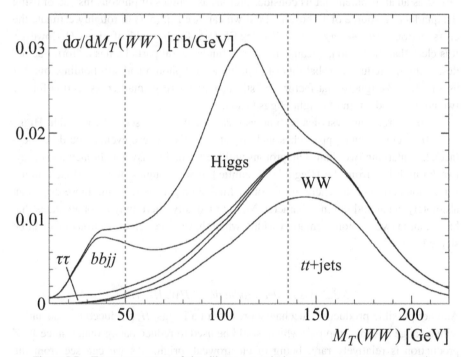

Figure 5.7   Transverse mass of the dilepton + missing transverse energy system in events with two detected tag jets. The Higgs resonant peak is evident above the continuum *WW* and other backgrounds, even for the "worst" case of a Higgs boson with 115 GeV mass ([5] – with permission).

The COMPHEP Feynman diagrams for production of a Higgs via *WW* fusion with subsequent decay into *WW* or *WW** (Higgs with masses below *WW* threshold that decay have one of the *W* "off mass shell" or virtual, which is indicated as *W**) are shown in Fig. 5.6. The transverse mass of the *WW** system where the *W* both decay into a lepton plus neutrino is shown in Fig. 5.7 for a Higgs mass of 115 GeV, which is the final LEP II upper mass range. For masses greater than this but less than ∼200 GeV, the situation in regards to the signal to background and size of the cross section is even more favorable.

Figure 5.8   Feynman diagrams from COMPHEP on the associated production of Higgs particles and gauge bosons. The process is Drell–Yan production where the off shell W or Z subsequently radiates a Higgs boson.

Indeed, the WW fusion process, with detected tag jets, is an important discovery mode for the Higgs search. It is also important to notice that this process depends only on the Higgs coupling to the gauge bosons, HWW, so that this coupling can be isolated and measured.

Just as an amusement, let us consider protons as sources of photons instead of using the quarks as a source of W bosons. First we can use Eq. (5.7) to roughly estimate the cross section for $pp \to pp\gamma\gamma \to ppH$, using Eq. (5.5) for the width of $\gamma\gamma \to H$. However, it is clear that the proton, because it is not a fundamental particle, has a "form factor" describing the reduced probability of emitting a hard photon and still holding together as a proton. We ignore that factor, and still arrive at a very small cross section for the two photon production of a light Higgs boson.

A more careful analysis leads to an estimate for the cross section for a light Higgs of $\sim 10^{-39}$ cm^2, which approaches viability at the LHC. These events would be spectacular, containing two final state protons with very small transverse momenta and, say, two b quark jets from the Higgs decay emitted at wide angles, with ~ 60 GeV transverse momentum each. There are no other final state particles, giving these events an absolutely clean and unique character. More practically, there may be other states that have a large two photon formation width, which have larger cross sections that can be studied.

5.3.3 Associated production – HW, HZ, Htt

Another possible production mechanism results in a Higgs, H, produced in association with a gauge boson or top pair which would be used to reduce backgrounds since W, Z production is relatively rare, being of electroweak origin. As we can see from the COMPHEP Feynman diagrams, Fig. 5.8, the production mechanism involves Drell–Yan formation of a virtual W or Z with subsequent Higgs Bremsstrahlung. A measurement of this process would clearly probe the Higgs coupling to gauge bosons. The cross section is, however, 10 to 100 times less than the main production mechanism, gluon–gluon fusion (see Figure 5.3).

The Higgs plus gauge boson production process is advantageous because it improves the signal to background in the case of a low mass Higgs boson. However, the cross section falls rapidly with Higgs mass limiting the utility of this mechanism to low Higgs masses. Production by quarks is also more advantageous at lower CM energies. The favored Higgs search strategy at CDF and D0 will be to use associated production with Higgs decay into b quark pairs. The results of a Monte Carlo simulation for Higgs signal

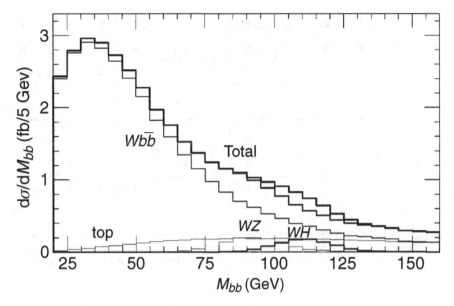

Figure 5.9 Mass distribution for *b* quark pairs due to *WH*, *WZ*, and top pairs decaying into *W* + *b*, and continuum *W* + *b* pair production. The *W* decays into electron plus neutrino and the Higgs mass is 110 GeV. The model is for 2 TeV CM energy, proton–antiproton collisions at the Fermilab Tevatron where the production of *WH* by quarks is enhanced with respect to gluonic production ([6] – with permission).

and backgrounds due to the continuum production of $W b\bar{b}$ pairs and other processes is shown in Fig. 5.9.

At the LHC, at the appropriate *x* values for light Higgs masses there is an enormous population of gluons which makes this strategy rather more difficult, and it will not be considered further here.

Another process that has great promise to reduce backgrounds is the production of a Higgs boson in conjunction with a pair of top quarks, which exploits the strong coupling of the Higgs to the top quark. The COMPHEP Feynman diagrams for gluon–gluon production of that final state are shown in Fig. 5.10. The cross section is rather large (see Fig. 5.3) because the couplings are *Htt*, and the large top mass means that this coupling is quite strong. A measurement of the rate for this process will help us probe the SM prediction for the top quark couplings to the Higgs.

The $Z + t + \bar{t}$ final state gives us a convenient "control" sample because the Feynman diagrams are identical to those for $H + t + \bar{t}$, the production cross sections are similar, and the clean detection of the *Z* in the dilepton decay mode is well established. The cross section for the QCD background process is shown in Fig. 5.11 as given by COMPHEP.

The cross section for the production of $H + t + \bar{t}$ for a 120 GeV Higgs mass is ∼0.3 pb (see Fig. 5.3). Assume all light Higgs decay solely into *b* quark pairs. The calorimetric mass resolution for reconstruction of the resonance from measurement of the two *b* jets is expected to be dM/M ∼0.06 (see Chapter 2). Assuming the entire signal is contained

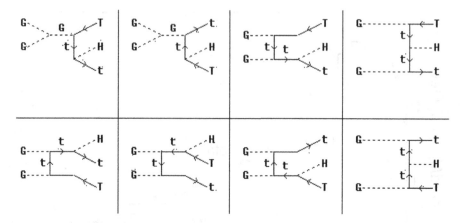

Figure 5.10 COMPHEP Feynman diagrams for the process $g + g \rightarrow H + t + \bar{t}$. The relevant Higgs couplings are to the top–antitop quark pair. Diagrams for the similar process where the H is replaced by a Z gauge boson yield a comparable cross section.

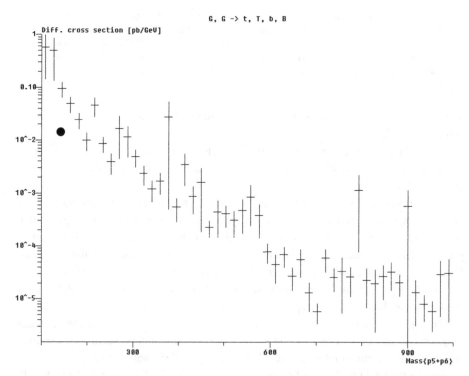

Figure 5.11 Cross section for $g + g \rightarrow t + t + b + b$ at the LHC as a function of the b pair mass, M, as given by the COMPHEP Monte Carlo program. The expected Higgs signal for the b pair decay mode and a 120 GeV Higgs is also shown as a dot.

in $3 \times dM$ or ~ 22 GeV (ignoring the effect of the very small natural width), the signal is a resonant "bump" of height ~ 0.014 pb/GeV. The signal to background ratio, looking at Fig. 5.11, is then reasonably favorable, $S/B \sim 1/6$.

As we will see, the QCD background for gluon–gluon production of a Higgs boson, which then decays into b quark pairs, is insurmountable. Using the much-improved signal/background ratio available in the $H + t + \bar{t}$ production process, we can attack this difficulty and thus can determine the Higgs branching fraction into b quark pairs. Clearly, that is a crucial measurement since it tests the SM prediction for the Yukawa coupling of the Higgs boson to the fermions' mass.

5.3.4 Pair production of Higgs

The assumed interaction potential energy for the Higgs field is $V(\phi) = \mu^2 \, \phi^2 + \lambda \phi^4$, which we showed in Chapter 1 and Appendix A. The parameter μ has the dimensions of mass, while λ is dimensionless. The vacuum exists at the minimum of this potential with vacuum field $\langle \phi \rangle = \sqrt{-\mu^2/2\lambda}$. Expanding around the minimum, $\phi \sim \langle \phi \rangle + \phi_H$, we collect terms with the same powers of the fields. The terms quadratic and higher in the Higgs excitation ϕ_H are (ignoring numerical coefficients):

$$V(\phi_H) \sim \lambda \left[\langle \phi \rangle^2 \phi_H^2 + \langle \phi \rangle \phi_H^3 + \phi_H^4 \right]. \tag{5.9}$$

The first term is easily identified, see Appendix A, as an effective mass term, with $M_H = \sqrt{2\lambda} \langle \phi \rangle$. Thus, the Higgs acquires a mass, but the numerical value is not predicted because it depends on the unknown parameter λ. For this reason, we need to adopt a wide ranging and flexible search strategy, one which has a good chance of success and covers a mass range from the lowest experimentally allowed value set by existing LEP searches, ~ 115 GeV, to the highest values, set by the point at which the weak interactions become strong, ~ 1.7 TeV.

The other terms correspond to self-couplings of the Higgs. The triplet term has an effective coupling $\sim \lambda \langle \phi \rangle \sim \sqrt{\lambda} M_H$, while the quartic term has, as expected, a dimensionless coupling $\sim \lambda$. Therefore the Higgs couplings in the SM are: to gauge bosons $\sim g_W M_W$, to fermions, $\sim g_W \, (m_f/M_W)$, triple self-couplings, $\sim \sqrt{\lambda} M_H$, and quartic self-couplings, $\sim \lambda$. Once the Higgs mass is measured we know λ and the self-couplings are completely specified if the SM is the correct description of Nature. Therefore, a measurement of these self-couplings would be a very important check of the SM.

The most important Feynman diagrams for Higgs pair production are shown in Fig. 5.12. The cross section depends on the triple coupling of the Higgs. This situation is similar to the case of gauge boson pair production, which depends on triple gauge couplings.

The cross section at the LHC for Higgs pairs is shown in Fig. 5.13 as a function of the Higgs mass. The cross section level is quite low, ~ 20 fb for a light Higgs mass. At design luminosity this means 2000 Higgs pairs produced in one year at the LHC.

Experimentally clean signatures with a high enough branching ratio so that the *HH* signal can be observed at the LHC seem to be very difficult to arrange. For example, at

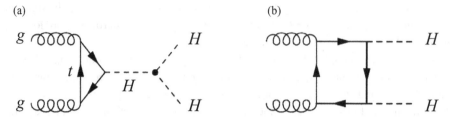

Figure 5.12   Feynman diagrams, (a) triple Higgs coupling and (b) "box" diagram with top quarks radiating a Higgs boson twice, which are the most important for the pair production of Higgs bosons at the LHC ([6] – with permission).

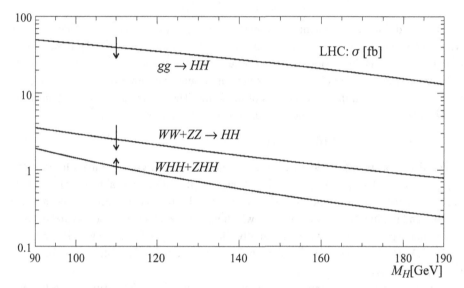

Figure 5.13   Cross section for the production of Higgs pairs at the LHC as a function of the mass of the Higgs boson ([7] – with permission). The major process is gluon–gluon fusion, as is the case for single H production.

low Higgs mass, the decay into b quark pairs dominates. Thus there are $\sim$2000 Higgs pairs decaying into four b quarks. However, the background of four b events from QCD sources is overwhelming. If we use one Higgs decay to b pairs and the other decay into W^*W we must pay in event rate for the branching ratio into W^*W and the subsequent leptonic W decay branching fractions.

At present, there is no good search strategy worked out for detecting Higgs pairs at the LHC. This is a serious shortcoming, since the SM makes an unambiguous prediction about Higgs self-couplings, which must be checked. Work to find a strategy to measure this process continues. In particular, an upgrade of the LHC with ten times more luminosity is being contemplated. This increase in luminosity might allow us to exploit triggerable decay modes of a light Higgs, such as $H \rightarrow W^* + W$ with a subsequent

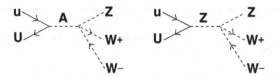

Figure 5.14   COMPHEP Feynman diagrams for the Drell–Yan production of three gauge bosons. Only those diagrams containing quartic couplings are shown.

leptonic decay of one W, $W^- \rightarrow e^- + \bar{\nu}_e$, and quark–antiquark decay of the other W, such as $W^+ \rightarrow u + \bar{d}$.

5.3.5 Triple gauge boson production

Although not strictly part of the Higgs search, a measurement of triple gauge boson production is a probe of the predicted SM quartic couplings of gauge bosons. As we saw in Chapter 4, the presently available data from CDF and D0 contain only a small number of gauge boson pairs and the LEP data contain only a few $WW\gamma$ events (Chapter 1). The increased luminosity and CM energy of the LHC will make the production of three gauge bosons experimentally accessible.

At the LHC the cross section to weakly produce W pairs directly is ~ 100 pb. The strong production of top pairs (thus W pairs) has a cross section of ~ 800 pb. The cross section for the production of the WWZ final state is ~ 3 pb, not reduced too greatly, $\sim \alpha_W$, below the weak W pair cross section because the CM energy is so much larger than the sum of the masses in the final state. Therefore, at the LHC the SM prediction for quartic couplings can be confronted very directly. For example for WWZ with both W decaying into leptons and Z decaying also into electrons or muon pairs $(\ell^+ + \nu_\ell + \ell^- + \bar{\nu}_\ell + \ell^+ + \ell^-)$, in one year there are ~ 1000 events assuming full efficiency for triggering and reconstruction. Note also that the gauge boson pairs from WW fusion (see Fig. 5.6) at high WW mass probe the strength of the quartic coupling predicted in the SM.

5.4 Higgs branching ratios and search strategy

We now put together what we have learned about the coupling of the Higgs boson to quarks, leptons, and gauge bosons and what we have learned about the cross section for production of the Higgs by means of different production mechanisms. Depending on the rarity of the final state with respect to the specific backgrounds existing for that particular final state, different production mechanisms may be needed on a case-by-case basis if we are to fashion a successful search strategy. That strategy is very dependent on the unknown Higgs mass. For example, a basic issue is whether the Higgs mass is sufficiently large to use the relatively straightforward ZZ final state or not. Our goal is to fashion a search strategy that can both discover the Higgs and also inform us about its coupling to leptons, quarks, and bosons independent of what the Higgs mass turns out to be.

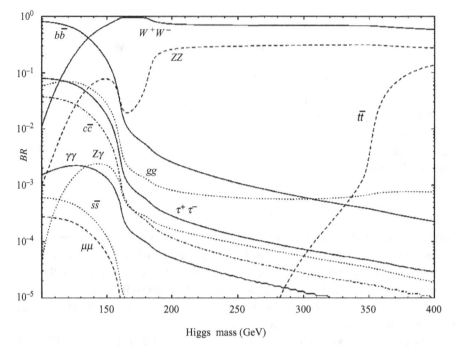

Figure 5.15   Branching fractions of the Higgs boson as a function of the Higgs boson mass ([8] – with permission).

Let us look at the branching fractions of a Higgs boson into different final states as a function of the Higgs mass. If the decay width to a final state i is Γ_i, then the total decay width Γ is $\sum \Gamma_i$ and the branching ratio is $B_i = \Gamma_i/\Gamma$. The branching ratios for the Higgs boson as a function of Higgs mass are shown in Fig. 5.15. The rapid variation with Higgs mass indicates the need to fashion a comprehensive search plan.

The Higgs width is very small below the WW "threshold." The heaviest available quark pair, $b\bar{b}$, dominates below WW "threshold" at a mass of $\sim$160 GeV. The charm pair branching ratio is estimated to be $B(c\bar{c}) \sim (m_c/m_b)^2 \, B(b\bar{b}) \sim (1.2 \text{ GeV}/4.5 \text{ GeV})^2 \sim 0.1$. The heaviest accessible lepton pair, τ, has a width reduced by $\sim$9 relative to the b pair width because of the coupling to mass squared, $(1.74 \text{ GeV}/4.5 \text{ GeV})^2$, and by a $1/3$ color factor, leading to a rough estimate of the branching fraction of $1/27 = 0.037$.

Our previous estimates of 6 MeV for the b quark width and 1.16 keV for the photon width lead to a light Higgs two photon branching ratio estimate of 0.000 19. The exact calculation yields a result about ten times larger, indicating that the W in the loop which we ignored makes a large contribution. The gg width estimate was 0.25 MeV or a gg branching ratio of 0.04. Comparing these "back of the envelope" estimates with the exact results shown in Fig. 5.15 we conclude that we roughly understand the most important decay modes for a low mass Higgs boson.

Decay widths for the Higgs are easily generated by COMPHEP. In Fig. 5.16 the b quark pair width is displayed. Note the linear behavior with Higgs mass, and the fact that

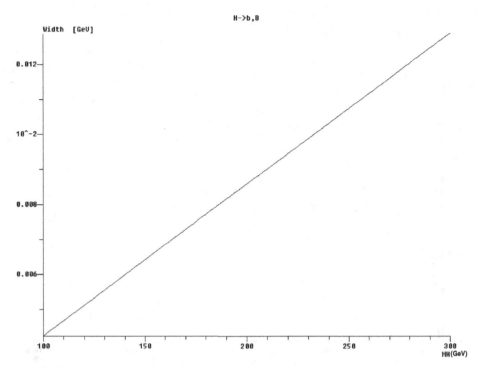

Figure 5.16   Decay width into *b* pairs is generated by COMPHEP as a function of Higgs mass *b* pairs.

a 6 MeV decay width into *b* pairs for a 150 GeV Higgs mass is confirmed. The width to τ pairs is also linear with a scale set by a width ~ 0.35 MeV at the same Higgs mass. The top pair width shows the threshold behavior, β^3, which is explained in section 5.4.1. The top pair width, because the mass is so large, can be substantial (see Fig. 5.15). However, there is a severe strongly produced, or "QCD background", of top pairs, as illustrated in the cross section estimates given in Chapter 4. Therefore top pairs as a way to detect the Higgs will not be considered further.

Representative decay widths for a Higgs mass of 250 GeV are given below.

$$\Gamma(H \to b\bar{b}) = 9.5\,\text{MeV},$$
$$\Gamma(H \to \tau\bar{\tau}) = 0.5\,\text{MeV}. \tag{5.10}$$

What about "below threshold" decays? As we mentioned in Chapter 1, below ZZ "threshold" there is a Zl^+l^- mode with an "off shell Z," conventionally called ZZ*. The decay width, $\Gamma_Z \sim 2.5$ GeV, and the Breit–Wigner resonant mass distribution, $d\sigma/dM \sim (\Gamma/2)^2/[(M - M_o)^2 + (\Gamma/2)^2]$ means that the ZZ* decay rate is suppressed by a factor of $\sim[(\Gamma_Z/2)/(M - M_Z)]^2$ with respect to ZZ decays as the l^+l^- mass goes off the resonant mass from M_Z to M. Therefore, going from a ZZ decay mode at 180 GeV, with a decay width of 0.3 GeV, to a decay rate at 150 GeV for ZZ*, we can expect an approximate decay width of 300 MeV(1.25 GeV/30 GeV)^2 ~ 0.5 MeV and a WW* width ~ 600 MeV(1.0 GeV/10 GeV)$^2 = 6$ MeV which is approximately the *b* pair width. Indeed, the

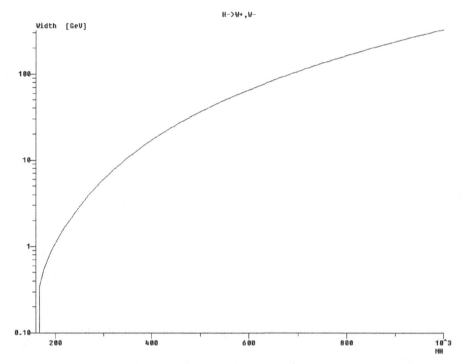

Figure 5.17   Decay width generated by COMPHEP as a function of Higgs mass for *W* boson pairs.

below threshold branching fractions for *WW** and *ZZ** shown in Fig. 5.15 are roughly of that magnitude.

The width above *WW* threshold generated in COMPHEP is shown in Fig. 5.17. The corresponding *ZZ* width is half that for *WW*, as expected from Eq. (5.4), and the rise with Higgs mass is shifted by about twice the *W, Z* mass difference, or ZO GeV.

After a steep rise from threshold, the $\Gamma \sim M^3$ behavior we expect is clearly seen in Fig. 5.17. Widths into *W, Z,* and top pairs at 600 GeV Higgs mass are presented in Eq. (5.11). The width to mass ratio at 600 GeV is quite large, $\Gamma_H/M_H \sim 0.21$. It extrapolates to $\Gamma_H/M_H \sim 1$ when $M_H \sim 1.7$ TeV.

$$\Gamma(H \rightarrow WW) = 70\,\text{GeV},$$
$$\Gamma(H \rightarrow ZZ) = 35\,\text{GeV},$$
$$\Gamma(H \rightarrow t\bar{t}) = 20\,\text{GeV}. \tag{5.11}$$

Of the decay modes mentioned so far, the $H \rightarrow \gamma\gamma$ decay mode is a clean method to search for low mass Higgs using direct Higgs production. The *b* pair and τ pair decay modes are also accessible at low mass if the *ttH* (associated production) and *qqH* (*WW* fusion with tag jets) production mechanisms are employed respectively to improve the signal to noise ratios. Above an effective threshold for *ZZ** at ~ 170 GeV Higgs mass the

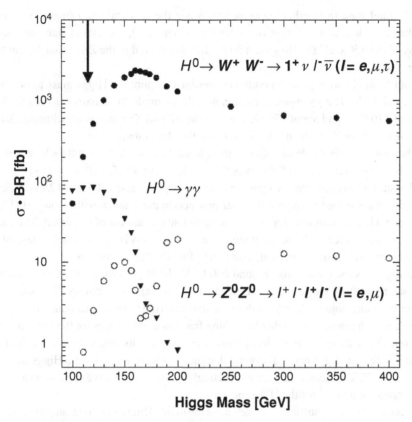

Figure 5.18 Cross section times decay branching ratio as a function of Higgs mass for Higgs decays into vector boson pairs at the LHC. The present LEP II limit is indicated by an arrow ([1] – with permission).

four lepton mode is clean and is the process of choice since direct Higgs production is usable. The *WW** decay to two leptons and two neutrinos does not have a sharp transverse mass peak due to loss of information about the longitudinal momentum of the neutrinos (Fig. 5.7). Nevertheless, with the use of tag jets to signal *WW* fusion production, the $H \to W + W^* \to (l^+ + \nu_\ell) + (l^- + \bar{\nu}_\ell)$ decay scheme is a major "discovery mode" for Higgs particles with mass < 200 GeV. We expect that the branching fraction to *W*W* will be the largest Higgs mode for Higgs mass above about 140 GeV (see Fig. 5.15).

The production cross section (Fig. 5.3) times decay branching ratio (Fig. 5.15) is shown in Fig. 5.18 for *WW*, *ZZ*, and $\gamma\gamma$ decay modes assuming *g–g* production and leptonic decays of the *W* and *Z* bosons. For Higgs masses from 100 to 400 GeV the detected cross section times branching ratio into the two photon or four charged lepton final state is always >10 fb. This means that at least 1000 Higgs events are produced and decay into a clean, detectable final state in one year of LHC data taking at design luminosity. Since the four charged lepton final state is well measured by tracking detectors, the *Z* resonances will appear as prominent features, allowing us to cleanly extract the ZZ final state from other backgrounds. Since the ZZ continuum final state is only produced with a cross section of $\sim$ pb (see Chapter 3), we expect that for masses above about 150 GeV the Higgs can be readily discovered in the ZZ to four lepton final state.

Data from electron–positron colliders presently require the Higgs mass to be above about 110 GeV. The $\gamma\gamma$ mode is the cleanest decay mode for masses between the LEP limit of 110 GeV and about 150 GeV where the ZZ and ZZ^* modes are difficult. Above 150 GeV, ZZ^* or ZZ is the mode of choice in the Higgs search.

The WW^* or WW mode can also be used from $\sim$120 GeV to $\sim$ 200 GeV, where the large branching ratio into WW^* makes this mode attractive. The $(l^+ + \nu_\ell) + (l^- + \bar{\nu}_\ell)$ final state rate exceeds the two photon rate for Higgs masses above $\sim$125 GeV, even though we are forced to require the qqH production mechanism with a rate $\sim$1/10 the rate shown in Fig. 5.18 in order to achieve sufficient cleanliness of the signal. This mode is, therefore, a potential "discovery mode" in the low mass Higgs region, because of the large event rate, perhaps the most promising of all the "clean" modes.

At high masses, greater than around 600 GeV, the Higgs cross section falls so much that we run out of the statistics we need for a compelling discovery. The somewhat dirtier but more copious decay modes of Z into neutrino pairs and jet pairs are required at high mass because their higher branching fraction compensates for the reduced cross section. The addition of those decay modes extends the discovery "reach" of the LHC up to $\sim$1 TeV in Higgs mass. Theoretical arguments tend to require the Higgs mass not to exceed 1 TeV although they are not particularly crisp, so that we can cover the entire mass range "allowed" for the Higgs boson.

This introduction is sufficient to sketch out the main elements of the strategy to discover the Higgs, whatever its mass. We now begin a more detailed discussion of the search strategy for particular decay modes and explain why a given strategy applies only over a limited range of Higgs mass.

5.4.1 $b\bar{b}$

In general a quark–antiquark pair in a state of total spin S and angular momentum L has a parity, P, and charge conjugation quantum number, C, where $P = (-1)^{L+1}$, $C = (-1)^{L+S}$. Therefore, the $J^{PC} = 0^{++}$ Higgs boson decays into P wave, $L = 1$ pairs. This, in turn, leads to a $\beta^{(2L+1)} = \beta^3$ threshold behavior for the decay width, where β is the quark velocity with respect to c in the CM frame.

We assume that the dijet invariant mass is calorimetrically reconstructed. For a Higgs mass of 120 GeV, the cross section is 30 pb. With a three standard deviation ($\pm$ 1.5 σ) signal region of $b\bar{b}$ mass, $\Delta M = 22$ GeV, set by the experimental resolution of the calorimetry the signal appears as a $\sigma/\Delta M = 30$ pb/22 GeV $= 1.4$ pb/GeV, resonant "bump" above the continuum cross section for the QCD production of b quark pairs (we assume that the 120 GeV Higgs b pair branching fraction is 1).

The COMPHEP Feynman diagrams for the QCD production of continuum b quark pairs are shown in Fig 5.19. The predicted cross section at the LHC is shown in Fig. 5.20.

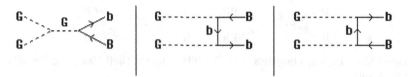

Figure 5.19   COMPHEP Feynman diagrams for the process $g + g \rightarrow +b + \bar{b}$.

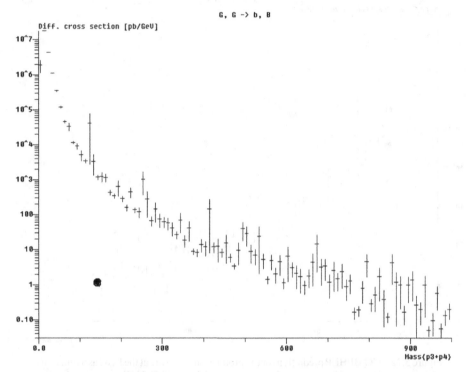

Figure 5.20   COMPHEP prediction for the production of b quark pairs at the LHC as a function of the quark pair mass. The dot represents the Higgs signal level for a 120 GeV mass Higgs.

The signal is also indicated in Fig. 5.20. It is swamped by a factor $\sim$1000. It is for this reason that we were forced to consider *Htt* production with subsequent $H \rightarrow b + \bar{b}$ decay, where the signal to background ratio is much more favorable (Fig. 5.11). Using the associated production mechanism, we can extract the cross section times $b\bar{b}$ branching ratio for light Higgs bosons and thus measure the Higgs coupling to b quarks.

5.4.2 $\tau^+\tau^-$

Another experimentally accessible decay mode for a light Higgs is that into τ lepton pairs. The COMPHEP Feynman diagrams for the production of the background continuum of τ pairs are displayed in Fig. 5.21. Basically, this background comes from Drell–Yan production of a virtual Z or photon which then decays into τ pairs.

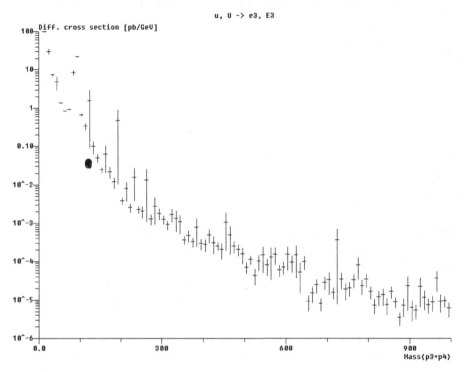

Figure 5.21   Feynman diagrams from COMPHEP for the Drell–Yan production of τ lepton pairs.

Figure 5.22   COMPHEP prediction for the production of τ pairs at the LHC as a function of the mass of the pair. The expected resonant signal for a 120 GeV Higgs is also shown.

The estimate we make for the Higgs signal in τ pairs is similar to that which we made for *b* pairs. At 120 GeV, assuming a branching ratio of 1/27 (see Fig. 5.15), we expect a resonant signal of 0.052 pb/GeV in a mass range of ∼22 GeV about the central Higgs mass. The COMPHEP prediction for the background continuum mass distribution is given in Fig. 5.22.

Clearly, the signal to background, *S/B*, ratio in this case is of order one, which is quite favorable. That improvement with respect to *b* pairs occurs because the coupling for the background process is electroweak and not strong and because the initial state partons are quarks rather than the much more copious gluons. We expect that we can extract the Higgs branching fraction into τ pairs if the Higgs mass is low, near the minimum mass that is not already ruled out by LEP. That result can then be compared with the branching fraction into *b* pairs from *Htt* associated production. In the Standard Model all branching ratios are predicted once the Higgs mass is specified.

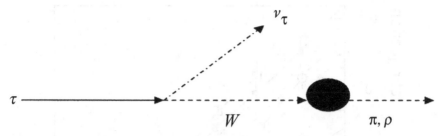

Figure 5.23   Schematic representation of the decay of a τ lepton. The final state contains a neutrino and a small number of charged particles, one in the case shown. The complicated "decay" of the virtual W into a quark pair and the subsequent quark "hadronization" into a pion or rho meson is indicated by a large dot.

However, the discussion of background so far applies only to direct τ pair production. There is also a large electroweak background due to W pair production with subsequent tau + neutrino decays, $W^- \to \tau^- + \bar{\nu}_\tau$, of both bosons. This electroweak background is also "irreducible," and differs from the signal process only in the presence of additional unobservable neutrinos in the final state. We assert that this difference allows us to control this source of background.

There are also tertiary sources of background. The QCD, or strong production, of top pairs leads to W and b pairs in the final state. The W pairs can then decay into tau + neutrino. This source of background can be reduced by "vetoing," or rejecting, events with extra jets. It turns out that the WW fusion mechanism, with visible tag jets, is needed to supply enough background rejection so that the τ pair in the final state from Higgs decay is visible above background.

We have so far assumed that "tau jets" can be selected with no background. This is not the case. A reducible background from QCD jets (e.g. gluons) exists that can also swamp the signal. We need a way to distinguish between QCD quark and gluon jets and tau jets. In order to understand how to do that, we must first look at the decay modes of the τ lepton. Since it is coupled to the W, the first step in tau decay is a virtual decay into a tau neutrino and a W. The W then virtually decays into quark and lepton pairs. The leptonic decays, $\tau^- \to \nu_\tau + \mu^- + \bar{\nu}_\mu, \nu_\tau + e^- + \bar{\nu}_e$, have small branching fractions which we ignore. For the quark decays of the virtual W, the particles in the final state are $\bar{u} + d$, which have the quark content of π^- or ρ^- mesons. Since the τ mass is only 1.74 GeV, it has a rather limited final state pion multiplicity. The τ hadronic decays are illustrated in Fig. 5.23.

The τ hadronic final state is, therefore, characterized by missing energy and a "narrow" jet normally containing only a single charged particle. This is rather different from a gluon jet, where the charged multiplicity is high and no neutrinos are directly emitted. Using these fundamental differences, the background from the strong QCD processes can be reduced sufficiently that the search strategy using τ pairs is a valid one.

As an example, in Fig. 5.24 we show the result of a Monte Carlo simulation of the rejection power against QCD jets as a function of the efficiency for τ jets. The τ jet is simply defined to be a "narrow" jet in (η, ϕ) space (see Chapter 2). Tracking multiplicity

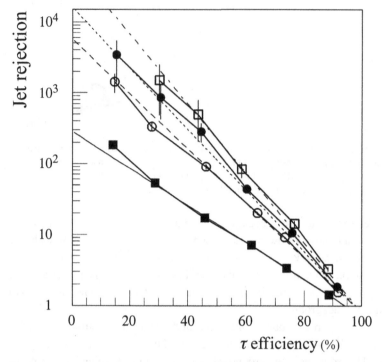

Figure 5.24   Plot of the τ efficiency as a function of the QCD jet rejection factor for different ranges of the transverse momentum ranging from 15 GeV (shaded boxes) to 130 GeV (open boxes). The QCD background rate is a function of the "narrowness" of the jet that forms the trigger (ATLAS figure – with permission).

is also used to require low charged multiplicity in the jet. At a reasonably high transverse momentum, a 50% efficiency for τ jets is retained while a hundred-fold rejection against QCD jets is achieved.

5.4.3 $\gamma\gamma$

The final branching mode of current experimental interest specific to low mass Higgs bosons is that into two photons. Basically, it has a small branching ratio but is experimentally quite clean. Recall that in Chapter 4 we looked at the experimental data on two photon production and compared them with a COMPHEP Born approximation prediction arising from the reaction $u + \bar{u} \rightarrow \gamma + \gamma$.

The resulting COMPHEP prediction for the continuum background of photon pairs at the LHC is shown in Fig. 5.25.

We expect the $\gamma\gamma$ mode background to be more favorable than the $b\bar{b}$ decay mode because the initial state probability is smaller, quarks instead of gluons, and the strong production (QCD) is reduced to electromagnetic (QED) coupling strength, similar to the situation with τ pairs. In addition, the mass resolution for electromagnetic calorimetry

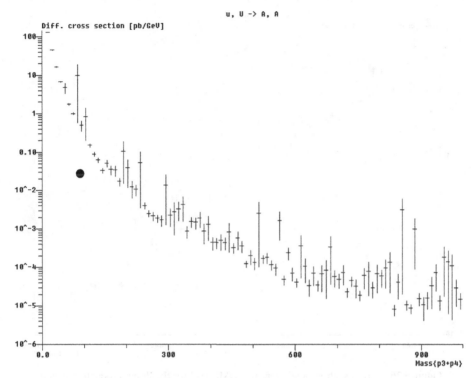

Figure 5.25   Cross section for photon pairs at the LHC as a function of the pair mass.

is about ten times better than for $b\bar{b}$ or τ pairs (see Chapter 2). Therefore, in the two photon case we can exploit the full rate of gluon–gluon fusion Higgs production and not be forced to use some lower rate process using associated Higgs production.

Taking a 120 GeV Higgs mass, a 2 GeV mass "window," and a branching ratio of 0.002 (Fig. 5.15), we expect a resonant signal in the two photon mass spectrum of (30 pb) (0.002)/2 GeV = 0.03 pb/GeV. The signal is still buried by a factor $\sim$10 in the background, so there is a premium on obtaining the best possible calorimetric energy resolution. Nevertheless, the signal has a clean signature, and in the mass range just above the present experimental Higgs mass limit, it will be a primary search strategy. A spin 1 particle cannot decay into two photons. That is a significant restriction on the quantum numbers, since fundamental bosons are not thought to exist in the SM with spin greater than 1.

There is also a reducible strong QCD background from neutral pions that decay into two photons. If these are not resolved, the strongly produced pion will be an additional large background to the photon. For a $\sim$100 GeV Higgs mass, the symmetric decay to a pair of photons implies that the photons have $\sim$50 GeV transverse momentum. These photons are mimicked by 50 GeV neutral pions (mass $M_\pi = 0.14$ GeV), which subsequently decay into photon–photon pairs with opening angle of about $(2M_\pi/M_H) \sim 0.003$ rad. If the electromagnetic calorimeter is placed at a transverse distance $r \sim 2$ m

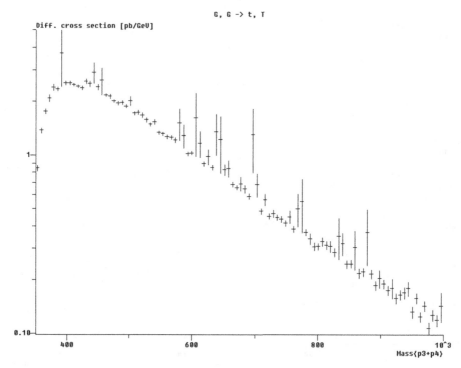

Figure 5.26 Cross section prediction by COMPHEP for the direct production of top pairs at the LHC as a function of the pair mass.

from the interaction point, the photons are separated by ∼0.6 cm at the point of impact on the calorimeter. Therefore, the "pixels" of the calorimetry (see Chapter 2) need to resolve clusters of electromagnetic energy with this scale of transverse separation. The LHC experiments have prepared for this challenge by employing small "pixels" in their highly segmented electromagnetic calorimeters.

5.4.4 *WW*→ $(\ell\nu)(\ell\nu)$

The production of top pairs proceeds by way of the same Feynman diagrams as the production of b quark pairs (same QCD dynamics because all quarks have the same color charge). Therefore, aside from kinematic effects due to the difference in mass, both processes should have the same cross section. The COMPHEP prediction for top pair production at the LHC is shown in Fig. 5.26 as a function of the quark pair mass. The cross section is indeed the same as that shown in Fig. 5.20 at high pair masses, above about 500 GeV.

These strongly produced top pairs lead to a large number of W pairs, since the top decays almost totally to $W + b$. These W pairs are a potential background if Higgs searches are performed using the *WW* or *W*W* final state. As we mentioned previously, the *WW* fusion mechanism with detected "tag" jets is used to make the Higgs decay

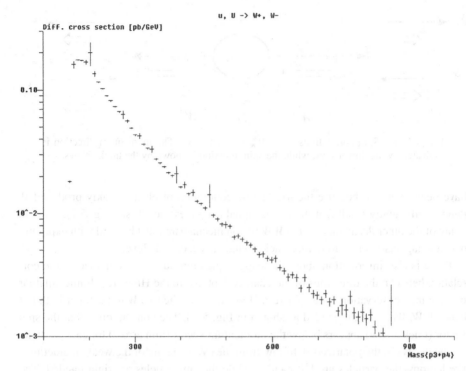

Figure 5.27   Cross section as a function of *W* pair mass for *WW* production in 14 TeV
p–p collisions at the LHC.

to *W* pairs accessible to experiment by increasing the signal to background value. The
existence of extra *b* quark jets in the QCD produced top pair background is exploited by
imposing a "veto" on additional jets in the event. Using this veto cut, the *W* pairs from
top pair decays can be strongly suppressed.

There are also weakly produced *W* pairs (see Chapter 4), which have a somewhat
smaller cross section. However, they are irreducible and form a continuum background
for Higgs searches in *WW* final states (see Fig. 5.7). The cross section as a function
of the *W* pair mass at the LHC is shown in Fig. 5.27. In Fig. 5.26 the cross section is
1 pb/GeV at a top pair mass of 600 GeV. This crudely compares with the 0.04 pb/GeV
mass distribution for a *WW* mass of 300 GeV in Fig. 5.27. Therefore, if we can reduce the
top pair background by a factor >25 by vetoing on extra jet activity, we can concentrate
on the weakly produced but irreducible *W* pair background.

Clearly it is of interest in itself to measure the production of *W* pairs. The cross section
depends on the $WW\gamma$ and *WWZ* couplings, which are specified in the SM. The improved
Tevatron experiments currently taking data will, however, make these measurements well
before the LHC starts taking data, so it is not exactly clear what will remain to be done
at the LHC.

The *W*W* and *WW* decays into two charged leptons and two neutrinos can be used to
discover the Higgs boson after the reducible *W* pairs arising from top pair production

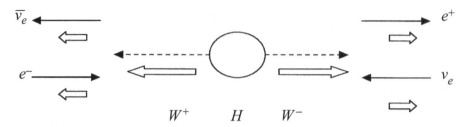

Figure 5.28   Spin correlations in $W_R W_R$ Higgs decays. The momentum direction is indicated by the thin arrows, while the spin direction is shown by the thick arrows.

have been removed, because the irreducible continuum of electroweakly produced W pairs is sufficiently small that the Higgs signal can be extracted (see Fig. 5.7). As with many of the other decay modes, the WW fusion mechanism must be used with explicitly detected tag jets in order to reduce backgrounds to acceptable levels.

There is also information about the Higgs' quantum numbers contained in the correlation between the directions of the charged leptons. In the Higgs rest frame, angular momentum conservation for a spin zero Higgs requires the two W to be either both left handed, $W_L W_L$, or right handed as shown in Fig. 5.28. The convention is that the spin vector (thick arrow) appears below the momentum vector (thin arrow) for each particle.

The decays of the polarized W follow from the (V-A) nature of the weak interactions. For leptons, the particles are left handed while the antiparticles are right handed. The overall effect is to make the charged leptons travel in the same direction. A vector resonance weakly decaying to W pairs would clearly not have the same charged lepton correlations. Therefore, a measurement of the momentum correlation of the charged lepton pair yields information on the spin of any observed resonance and may, in fact, be used to enhance the cleanliness of the signal.

5.4.5 $ZZ \rightarrow 4\ell$

The experimentally cleanest decay mode for the discovery of the Higgs boson is the decay to Z pairs with subsequent charged lepton decays of the Z. The $ZZ \rightarrow 4\,\ell$ branching ratio exceeds the two photon branching ratio for Higgs mass >150 GeV (see Fig. 5.18). The signal to background ratio in the ZZ final state is also much better. Therefore, for a Higgs mass > 150 GeV the final state of choice for overall cleanliness contains four charged leptons from ZZ or ZZ^*. The leptons are well measured in the tracker (see Chapter 2) and form a resonant state that is quite narrow ($\Gamma_Z/M_Z \sim 2.5$ GeV$/91$ GeV $= 0.027$). The Z pairs in turn have an excellent mass resolution. This decay mode is therefore called the "gold plated mode" for the Higgs search.

A schematic view of four electron and four muon events in the proposed CMS detector at the LHC is shown in Fig. 5.29. When low transverse momentum particles are not shown, as in the electron case, the event looks quite clean, containing only the four electrons and a recoil jet. In the muon case, the muon chambers themselves are protected

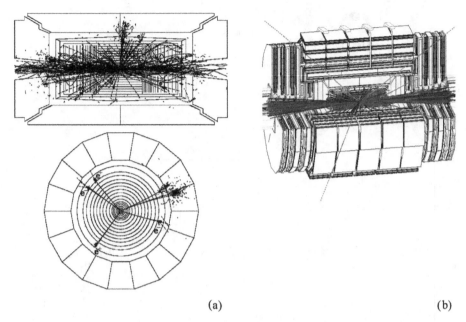

(a)   (b)

Figure 5.29   Plot of (a) a four electron and (b) a four muon Monte Carlo event arising from a Higgs decay into Z pairs in the CMS detector at the LHC. In the (r, ϕ) view only high P_T particles are plotted ([9] – with permission).

by the calorimatry and magnet yoke from the low momentum particles which are the vast majority of those produced. Thus, they largely see only the four isolated muons, again leading to a clean analysis.

For masses of 150, 300, and 600 GeV, the branching ratio into ZZ^* or ZZ is ~ 0.1, 1/3, and 1/3. The Higgs mass window due to the error on the tracking measurements of the lepton momenta is 2, 4, and 8 GeV while the Higgs natural width is 1.6, 13, and 105 GeV. The natural width dominates at high mass as expected. This leads to a three standard deviation Higgs mass window of 8, 45, and 330 GeV, or a cross section enhancement, $\sigma / \Delta M$, of 0.25, 0.074, and 0.002 pb/GeV in the ZZ mass spectrum.

The ZZ continuum background is due to Drell–Yan electroweak production of gauge pairs, similar to the WW electroweak background. The cross section for ZZ production is shown in Fig. 5.30 as calculated by COMPHEP for p–p production at 14 TeV. The expected signal is shown schematically for a 300 and 600 GeV Higgs boson. Clearly, the signal to background ratio is quite favorable in the four lepton final state because the background is due to a weak interaction production process. At higher masses, the search will become rather more difficult, simply because the Higgs becomes rather broad and the cross section falls rapidly with mass.

The integrated luminosity at design operation is 100 fb^{-1} per year. The cross section times branching ratio for a 600 GeV Higgs decaying into ZZ is $\sim$0.7 pb. The decay rate for Z into electron or muon pairs is 6.7%, or a 3 fb cross section into four leptons. Thus,

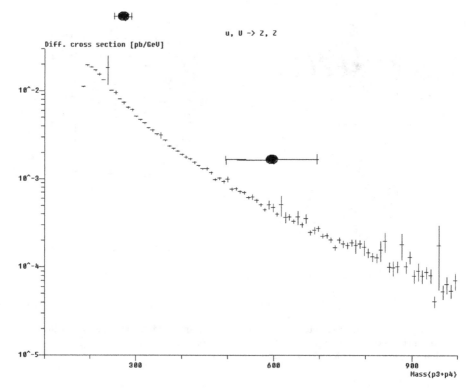

Figure 5.30 Cross section for *ZZ* production at the LHC as a function of the *ZZ* invariant mass. Also indicated are the signals expected for 300 GeV and 600 GeV Higgs bosons decaying into *Z* pairs.

with no background and perfect detection efficiency, we get 300 signal events or a 17 standard deviation signal.

CMS and ATLAS have made detailed studies and will see a ten standard deviation resonant signal in one year of operation for most of the mass range where *ZZ* or *ZZ** measurements are relevant (see Fig. 5.36). A complete Monte Carlo study of the CMS detector yields the mass plots, which are similar to that given in Fig. 5.30, shown in Fig. 5.31. The Higgs masses shown there are 300 and 500 GeV. The plots are for different total integrated luminosities, but for 500 GeV the design luminosity for one "LHC year" was assumed. Clearly, in all cases a distinct and highly significant resonant peak is observable.

For Higgs masses <200 GeV, we expect to be able to extract resonant signals into several final states, as we have demonstrated above. The resonant mass of the Higgs boson will be well measured at low mass where the natural width is dominated by detector resolutions. However, the natural width will not be well measured, because the experimental spectrum is not strongly dependent on the very narrow natural width. At high mass the natural width dominates over instrumental resolutions and can be measured at the 5% level.

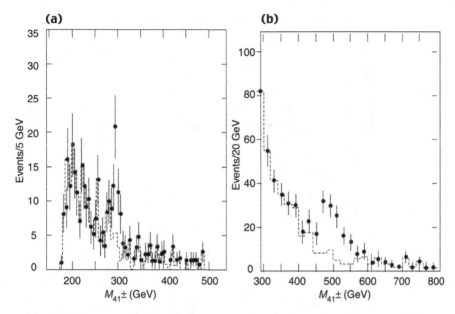

Figure 5.31   Monte Carlo predictions for the number of detected events in the ZZ to four lepton decay channel for Higgs masses of (a) 300 and (b) 500 GeV in the CMS detector ([8] – with permission). The number of events assumes an integrated luminosity of 20 and 100 fb^{-1} for (a) and (b) respectively.

A Monte Carlo study by the ATLAS collaboration of the expected error on some selected Higgs branching ratios is shown in Fig. 5.32. The γ partial width is determined by using the $\gamma\gamma$ final state. The partial width into b pairs uses Htt events. The WW fusion production of Higgs followed by the WW^* decay depends only on the HWW coupling, which allows us to clearly extract the W partial width. The ZZ branching fraction is found using the four lepton final state. The total width can be found at high Higgs mass by measuring the resonant line shape.

We see how the mass, total width, and some partial widths can be determined. What about quantum numbers? We can get some additional information on the spin, J, and parity, P, quantum numbers of the Higgs state from an analysis of the correlations among the ZZ decay products. It's amusing that, early in the study of high energy physics there was a "classic" pion parity experiment where the neutral spinless pion was observed to decay electromagnetically into two vector photons that then decay (rarely) into electron–positron pairs, $\pi^0 \rightarrow \gamma + \gamma \rightarrow e^+ + e^- + e^+ + e^-$. The analogy is to a neutral Higgs decaying electroweakly into two vector Z bosons and thence to four charged leptons.

For spin zero and positive parity the polarization vectors, $\vec{\varepsilon}$, of the photons are positively correlated and this is reflected in the alignment of the decay planes of the electrons. The opposite is true for the case of negative parity. The decay plane is that plane defined by the electron and positron momentum vectors. The parity is determined by looking at the correlation between the two decay planes. For $J^P = 0^+$, the decays have the decay

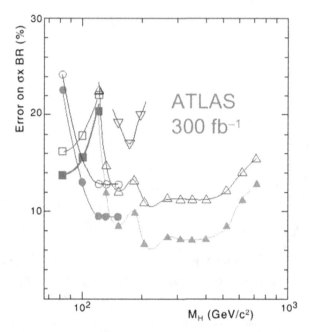

Figure 5.32   ATLAS study of the accuracy which can be obtained on the Higgs branching ratios for decays to two photons, circles, b quark pairs, squares, Z pairs, up diamonds,
and W pairs, down diamonds. The closed symbols indicate a 5% luminosity uncertainty,
while the open symbols are for 10%.

planes aligned, while for $J^P = 0^-$, the decay planes tend to be orthogonal:

$$\vec{\varepsilon}_1 \cdot \vec{\varepsilon}_2 \quad \text{for} \quad P = +,$$
$$\vec{\varepsilon}_1 \times \vec{\varepsilon}_2 \quad \text{for} \quad P = -. \tag{5.12}$$

The correlation of the lepton decay planes for spin zero and positive and negative parity
is shown in Fig. 5.33 for a 280 GeV Higgs mass, where the angle ϕ is the azimuthal
angle between the decay planes. Clearly, for positive parity the planes are preferentially
aligned, while for negative parity, they are orthogonal.

For a light Higgs mass, assume we have observed a two photon decay, so that the spin
is very likely zero. A scalar decaying into two vector gauge bosons is allowed in an *S*
wave, or zero orbital angular momentum state. The Z polarization can be longitudinal
(L) or transverse (T), since the Z has mass while the massless photon is transverse. The
decay distribution of the Z is $|Y_1^0|^2 \sim 1 + \cos^2 \theta$, $|Y_1^1|^2 \sim \sin^2 \theta$, for transverse and
longitudinal Z polarization, respectively, where the spherical harmonic is Y_l^m. The idea
is to fit the angular distribution for the fraction of Z_L and Z_T in the decays. There is an SM
prediction for the relative amount of transverse and longitudinal polarization of the Z.

$$\Gamma(H \to Z_T Z_T) / \Gamma(H \to Z_L Z_L) \sim (\delta^2/2)/(1 - \delta/2)^2,$$
$$\delta \equiv (2M_W/M_H)^2. \tag{5.13}$$

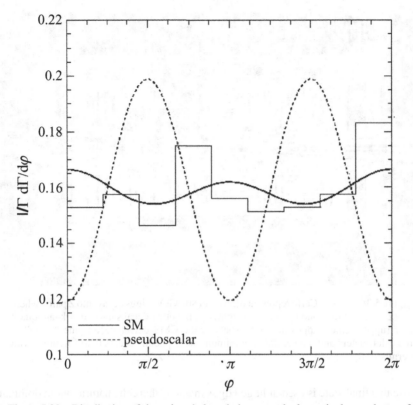

Figure 5.33 Distribution of the azimuthal angle between the leptonic decay planes of the two Z bosons in the decay $H \to ZZ$ of a 280 GeV Higgs in the case of scalar (solid line) and pseudoscalar (dotted line) bosons ([11] – with permission).

The two lepton decay of the Z serves as the analyzer of the Z polarization just as in the example of the two lepton decay of the photons. A fit to the decay angular distribution will determine the longitudinal and transverse components of the Z spin and thus test whether the Higgs quantum numbers are as predicted by the SM. At small Higgs mass, $M_H \sim 2M_W$, the T to L ratio is ~ 2, while at large Higgs mass, the Z will be completely longitudinally polarized.

5.4.6 $ZZ \to 2\ell + 2J$

For masses >600 GeV, larger branching ratio decay modes are needed due to rate limitations. We simply will not get enough events in a few years to be able to have a statistically compelling discovery. One possibility is to use the quark decays of the Z. The signal then appears in the two lepton, ℓ, plus two jet final state. The signal to background ratio is worse because the background from $Z +$ dijets due to QCD radiation in single Z processes is an added continuum contribution. The two jet mass resolution window for the other Z decay is also rather larger than the required leptonic Z decay mass window.

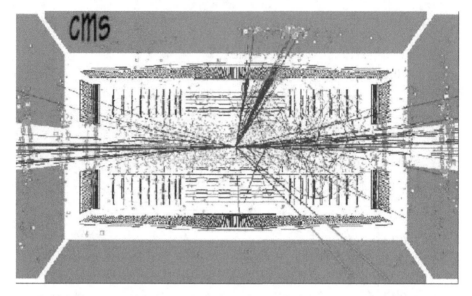

Figure 5.34   Monte Carlo representation of an 800 GeV Higgs decay into Z pairs, where one Z decays into a muon pair (lower right) and the other decays into a quark–antiquark pair (upper center) appearing as two jets in the CMS detector. Note the "noise" in the calorimeters and tracker due to minimum bias "pileup" events (CMS figure – with permission).

Because this final state is used at large Higgs masses where the natural width dominates over the detector resolution, using calorimetric mass determination is not, however, very costly in terms of sensitivity.

A Monte Carlo model of such a signal event is shown in Fig. 5.34 for a typical LHC detector. Suffice it to say that a Higgs signal can still be observed at large Higgs masses even with the enhanced background. Note the small opening angle for the jet pair arising from $Z \to q + \bar{q}$ decay. As discussed in Chapter 2, the angular segmentation of the hadronic calorimeter was chosen to resolve a Z decay into two distinct jets for Higgs masses up to 1 TeV. The event shown here illustrates why that choice of "pixel" size was made.

5.4.7 $ZZ \to 2\nu + 2J$

A still larger branching ratio can be used for the final state which occurs when one Z decays into a neutrino–antineutrino pair and the other decays into a quark–antiquark pair. In that case, we do not have the constraint that both pairs must be measured to have the resonant Z mass as we had for four leptons and, at rather worse mass resolution, for two leptons and two jets. Nevertheless, we can require a substantial missing energy and a large transverse mass for the first Z and a dijet mass of approximately the Z mass for the jets. There will be no invariant mass peak for the Higgs, but only a broad transverse mass enhancement. Therefore, the mass determination is not very good. Nevertheless, at these high masses the state is very broad anyway and the signal to noise ratio is still favorable since the backgrounds fall rapidly with increasing mass.

At some point we simply run out of events, even with the very high luminosity available at the LHC accelerator. The four jet final state is basically swamped by QCD strong production, and cannot be used in a Higgs search. Therefore, the falling cross section at high mass eventually makes the Higgs unobservable. This problem is exacerbated by the fact that the Higgs width is also rapidly increasing as the cube of the mass. The result is that the Higgs search terminates at a mass ~1 TeV for the LHC operating at design luminosity. If higher luminosities become available with "upgrades" to the LHC accelerator and to the detectors, the "mass reach" for the Higgs search will be extended beyond 1 TeV.

5.5 Luminosity and discovery limits

We have seen that the Higgs decay into *b* quark pairs is difficult to extract without the added background suppression achieved by using associated production with top pairs. This is not always the case. In the case that supersymmetry (SUSY) is a valid symmetry of Nature (see Chapter 6) a SUSY Higgs can have enhanced decay widths into *b* quark pairs. In some cases *WH* associated production can then be used to suppress backgrounds, allowing us to extract a resonant signal. The calorimetric energy resolution must be minimized as it directly defines the signal to background ratio. The results of a Monte Carlo study in this situation are shown in Fig. 5.35. The predicted experimental mass spectrum of signal plus background is shown for a particular set of SUSY parameters, which, in turn, influence the *b* quark branching fraction and decay width.

The width of 22 GeV, which is needed to contain the signal within the experimental mass resolution, is about five bins in Fig. 5.35. Hence, what we used in our previous estimations of calorimetric resolution is, if anything, an underestimate since there are other errors entering into the contribution of the mass resolution of a jet. Still, it is a good starting point.

The subject of how to search through the full parameter space of even some simple SUSY theories has a very extensive literature. The Higgs boson is no longer the simple object that we have assumed in the case of the SM. We continue, therefore, to concentrate on the simpler question of how we design a search for the SM Higgs boson using our accumulated knowledge obtained in Chapters 1–4. Some comments on searches for explicitly supersymmetric particles will follow in Chapter 6.

The figure of merit that is quoted in Fig. 5.35 is the significance or the number of signal events (S) divided by the square root of the number of background events B, or $S/\sqrt{B}$. In the limit of large numbers of events and small S/B ratio, this indicates the number of standard deviations by which the signal exceeds a statistical fluctuation of the background. The one standard deviation probability is 68%, two is 95%, and three is 99.77%. What is plotted in Fig. 5.36 is the significance, or $S/\sqrt{S+B}$, as a function of the Higgs boson mass for measurements using different final states during one year of operation at one third of design luminosity at the ATLAS detector. If the background is

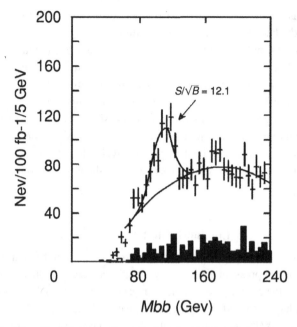

Figure 5.35   Mass distribution for *b* quark pairs for both background events and Higgs signal events for SUSY Higgs with a Higgs mass of 120 GeV ([12] – with permission).

much larger than the signal, then the significance becomes $S/\sqrt{B}$, as quoted in Fig. 5.35. If there is no background then the expression in this limit becomes $\sqrt{S}$, as we have already mentioned.

Basically, the CMS and ATLAS detectors are designed to discover the SM Higgs for all masses <1 TeV in four months of full luminosity operation. This is assumed to occur if a significance of about five standard deviations is achieved.

Clearly, the main final state that is used for Higgs discovery over a wide range of Higgs masses is ZZ or $ZZ^* \to 4\ell$. At high masses, larger branching ratio decay modes are needed and two leptons plus two jets or two charged leptons plus two neutrino final states are used. At low masses the two photon final state is used. The W^*W final state (in qqH), where both W decay into a lepton plus a neutrino, provides the largest sensitivity for Higgs masses, $\sim 2M_W$. Also at low mass, we have mentioned the $b + \bar{b}$ final state (in $H + t + \bar{t}$). The electron–positron collider LEP II has already set a mass limit ~ 110 GeV.

The LHC experiments and the LHC accelerator itself have been designed specifically to discover the Higgs boson that is hypothesized to exist in the SM. We expect that experiments at the LHC will discover the SM Higgs if it exists with a mass <1 TeV in the first year of full luminosity data taking. Depending on the mass of the Higgs boson, the width will also be determined, as will the branching fractions into a few final states. This information should help us to determine the interactions of the Higgs boson with quarks, leptons, and gauge bosons and compare them with the predictions of the SM. Once the Higgs mass is known, everything is predicted in the SM, so that any deviations

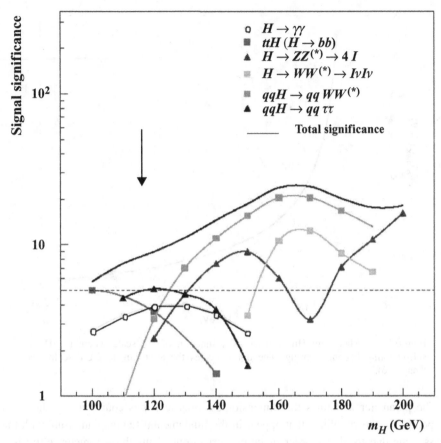

Figure 5.36 Expected significance at the ATLAS experiment as a function of the Higgs mass for 1 year of data taking at 1/3 of design luminosity (30 fb^{-1}). The LEP II limit is indicated by the arrow ([1] – with permission).

would allow us to conclude that new physics is making an appearance at this newly available high mass scale.

5.6 Lower limit on Higgs mass

We have argued that at high masses the Higgs boson ceases to exist as a distinct resonant state because the width is approximately equal to the mass at a mass of $\sim$1.7 TeV. There exists another argument, which indicates a still lower mass limit for the Higgs boson.

Recall from our discussion in Appendix A that the Higgs potential is $V(\phi) = \mu^2 \, \phi^2 + \lambda \, \phi^4$. There is a minimum at $\langle\phi\rangle$ that is non-zero, causing "spontaneous symmetry breaking." We then expand about the potential energy minimum, $\phi = \langle\phi\rangle + \phi_H$, in order to examine the behavior of the field excitations – the Higgs quanta. The curvature of the potential gives the Higgs mass, since the mass term in the Lagrangian density appears as $-M^2\phi^2$ and $(1/2)\partial^2 V / \partial \phi^2 = -M^2$.

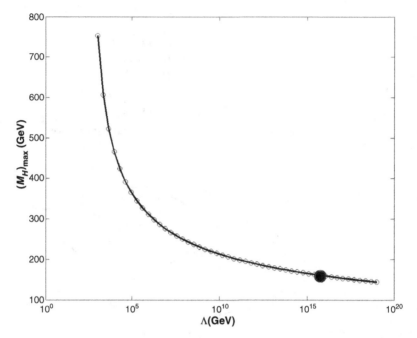

Figure 5.37   Maximum Higgs mass as a function of the scale where the Higgs self-coupling becomes strong. The dot indicates the approximate GUT scale (see Chapter 6).

The parameter λ defines the dimensionless quartic Higgs coupling. As shown in Appendix D, the couplings that appear in the fundamental Lagrangian "run" with the mass scale due to higher order quantum corrections. Thus the parameter λ too is a function of mass scale and varies logarithmically just as the SM coupling constants do. We simply assert that the behavior of $\lambda(Q^2)$ has the same "generic" behavior with mass scale as do the fine structure constants (see Appendix D), $\alpha(Q^2)$.

$$\lambda(Q^2) = \lambda(\langle\phi\rangle^2)/[1 - (3\lambda(\langle\phi\rangle^2)/8\pi^2)\ln(Q^2/2\langle\phi\rangle^2)],$$
$$1/\lambda(Q^2) = 1/\lambda(\langle\phi\rangle^2) - (3/8\pi^2)[\ln(Q^2/2\langle\phi\rangle^2)]. \tag{5.14}$$

The effective parameter, $\lambda(Q^2)$, increases with Q^2. If we require that $\lambda(Q^2)$ be well behaved from $\langle\phi\rangle = 176$ GeV up to a scale Λ, where $1/\lambda(\Lambda^2) = 0$ (strong Higgs self-coupling at the mass scale Λ), then $1/\lambda(\langle\phi\rangle^2) \sim 3/8\pi^2\ln(\Lambda^2/2\langle\phi\rangle^2)$.

Relating the parameter λ to the Higgs mass (see Appendix A), $M_H = \sqrt{2\lambda}\langle\phi\rangle$, we then have a constraint on the maximum value of the Higgs mass as a function of the mass scale where the Higgs quartic coupling constant diverges.

$$(M_H)_{max} \sim 4\pi\langle\phi\rangle/\sqrt{3\ln(\Lambda^2/2\langle\phi\rangle^2)}. \tag{5.15}$$

This constraint has no content unless we know at what scale the quartic couplings become strong. We show the scale dependence of the maximum Higgs mass in Fig 5.37.

If the scale is ~ 1 TeV then there is little new added to the 1.7 TeV limit we already have. On the other hand, if the scale is $\sim 10^{16}$ GeV (see Chapter 6, the SUSY grand unified scale), the limit is reduced to ~ 160 GeV. Numerically, $(M_H)_{max} \sim 1.26 \, \mathrm{TeV}/\sqrt{\ln(\Lambda^2/2\langle\phi\rangle^2)}$. If there is no new physics up to the scale of grand unification, then a light Higgs mass is favored.

Remember that at this mass the Higgs is narrow, has a reasonably large cross section, and has several accessible decay modes – b pairs (in ttH), τ pairs (in qqH), photon pairs, WW^* (in qqH), and ZZ^*. Thus, if we accept that the grand unified mass scale is relevant we expect the Higgs search to be very successful at the LHC and perhaps be accessible to CDF and D0 at the Fermilab Tevatron, depending on the performance of the accelerator. Recall that a low mass Higgs is also favored by the present precision electroweak data (see Chapter 4).

Exercises

1. Estimate the top pair cross section using Eq. (5.2). Compare with the COMPHEP results shown in Chapter 4.
2. For a luminosity of $10^{34}/\mathrm{cm}^2\,\mathrm{s}$, and a time interval for 1 year of $\sim 10^7$ s, estimate the number of 120 GeV Higgs, the number of W, and the total number of produced inelastic events at the LHC.
3. Show that the ratio of Higgs decay widths for WW and quark pairs is $\Gamma_{WW}/\Gamma_{q\bar{q}} \sim 1/6$ $(M_H/m_q)^2$. Therefore, above threshold for WW the gauge boson decays dominate.
4. Explicitly evaluate the widths into gluon pairs, b pairs, and photon pairs for a 150 GeV Higgs.
5. Find $(d\sigma/dy)_{y=0}$ numerically assuming gluon–gluon fusion production.
6. Explicitly evaluate the WW decay width for a Higgs of 150, 600, and 1200 GeV.
7. Work out the ratio of WW fusion to direct production of Higgs bosons.
8. Find the triple and quartic couplings of H by explicitly expanding the Higgs potential about the vacuum expectation value of the Higgs field.
9. Reproduce the distribution displayed in Fig. 5.20 using COMPHEP.
10. Reproduce the estimates of the signal shown in Fig. 5.10. Add a point for a 1000 GeV Higgs. Is it harder to find a 1 TeV Higgs from the point of view of S/B compared with a 300 GeV Higgs?
11. Evaluate Eq. (5.15) for two mass scales, one appropriate to electroweak symmetry breaking and one appropriate to grand unified theories (Chapter 6), $\Lambda \sim 10^3$, 10^{16} GeV.
12. Use COMPHEP to look at the "tag jet" process $u, d \to d, u, H$. Check the diagrams. Find the cross section for a 200 GeV Higgs mass and compare with the predictions given in Fig. 5.3.
13. Use COMPHEP to display the rapidity distribution of tag jets. Compare with the plot shown in Chapter 2.
14. Use COMPHEP to examine the process $u, U \to Z, H, H$. Is the cross section large enough to be observable at the LHC?

15. Use COMPHEP to explore the Higgs decays, $H \rightarrow 2*x$. Find the total width and branching fractions for several different masses. Remember COMPHEP has only "direct" decays.

References

1. Pauss, F., and Dittmar, M., ETHZ-IPP PR-98-09, hep-ex/9901018 (1999).
2. Spira, M. and Zerwas, P. M., *Lecture Notes in Physics* – 512, Berlin, Springer-Verlag (1997).
3. Braccini, S., arXiv:hep-ex0007010 (2000).
4. Kauer, N., Plehn, T., Rainwater, D., and Zeppenfeld, D., *Phys. Lett. B*, **503**, 113 (2001).
5. Baur, U., Ellis, R. K., and Zeppenfeld, D., Fermilab-Pub-00/297 (2000).
6. Djouadi, A., Kilian, W., Muhlleitner, M., and Zerwas, P. M., arXiv:hep-ph/9904287 (1999).
7. Djouadi, A., Kalinowski, J., DESY 97-079 (1997).
8. The CMS Collaboration, *CMS Technical Proposal*, CERN/LHCC 94-38 (1994).
9. The ATLAS Collaboration, *ATLAS – Detector and Physics Performance* – Technical Design Report, CERN/LHCC/99-15 (1999).
10. Choi, S. Y., Miller, D. J., Muhlleitner, M. M., and Zerwas, P. M., ArXiv:hep-ph/0210077 (Oct. 4, 2002).
11. Dittmar, M., ETHZ-IPP PR-99-06 and HEP-EX/9907042 (1999).

Further reading

ATLAS Collaboration, *Technical Proposal*, CERN/LHCC 94-43 (1994).
Carena, M. and J. Lykken, *Physics at Run II: The Supersymmetry/Higgs Workshop*, Fermilab-Pub-00/349.
CMS Collaboration, *Technical Proposal*, CERN/LHCC 94-38 (1994).
Donaldson, R. and J. Marx, *Physics of the Superconducting Supercollider – Snowmass 1986*, Singapore, World Scientific Publishing Company (1986).
Ellis, N. and T. Virdee, Experimental challenges in high-luminosity collider physics, *Ann. Rev. of Nucl. Part. Sci.* (1994).
Gunion, J., H. Haber, G. Kane, and S. Dawson, *The Higgs Hunters Guide*, Addison-Wesley Publishing Co. (1990).

6

SUSY and open questions in HEP

Something is happening and you don't know what it is, do you, Mr. Jones?

Bob Dylan

Toto, I've a feeling we're not in Kansas anymore.

Judy Garland

In the first five chapters we have focused rather tightly on the first two questions of the dozen raised at the end of Chapter 1. Those questions had to do with the spontaneous breaking of electroweak symmetry, which was assumed to be due to the vacuum expectation value of the Higgs field. That field gives the W and Z (and photon) a specified mass, $M_Z = M_W/\cos\theta_W$. It also gives masses to all the fermions of the SM via Yukawa couplings, but with unspecified values.

In addition, the SM predicts all the interactions of the Higgs once the mass is known. Since the mass is limited from below by experimental searches at LEP II to be $>110\,\text{GeV}$, and from above by general considerations to be $<1\,\text{TeV}$, we could map out a search strategy for the SM Higgs that almost guaranteed success at the LHC, assuming that this particle actually exists. Indeed the LHC and its experimental facilities are being constructed precisely for this purpose.

For a known Higgs mass, the width is predicted and can be compared with experimental data. We also need to measure the production cross section, both single and associated (H produced in association with W, Z, top pair). That will inform on the couplings of the Higgs to gluons, top quarks, and gauge pairs. We need to measure as many decay branching fractions as possible. Those data will tell us if the Higgs couples to the fermion mass as predicted in the SM. If the Higgs is heavier, the predicted coupling to gauge boson pairs must also be verified.

If it is a possible measurement, Higgs boson pair production will tell us about the triple self-coupling of the Higgs bosons. Observation of the decay to two photons would rule out a $J=1$ Higgs state. The angular distribution of the gauge pairs in Higgs decays, if kinematically available, allows us to determine the quantum numbers of the Higgs parent near the threshold for gauge pair decays. All this systematic study will allow experimenters at the LHC to determine whether a newly discovered resonant state at a given mass has some or all of the predicted properties of the Higgs boson specified in the SM.

After the Higgs boson (or something else which explains the W and Z mass) is discovered, there remain the other ten questions raised at the end of Chapter 1. As we look

189

into the outstanding questions for high energy physics, we will see that their explication might lead to additional experimental signatures that will also be closely examined at the LHC. These issues clearly go beyond the SM. Most high energy physicists think it unlikely that the search for the Higgs outlined in Chapter 5 will result in the discovery of a single resonant state at the LHC with all the properties of a fundamental scalar field. That judgment can only be tested experimentally. However, it is largely based on issues of taste. The plethora of arbitrary parameters that exist in the SM and the fact that the SM is not stable under quantum radiative corrections arising from the existence of a large Grand Unified Theory (GUT) or Planck mass scale argue that the SM is not a fundamental theory but an incomplete and therefore effective one.

3 – Why are there three and only three light "generations"?

6.1 Generations

The SM is widely felt to be incomplete because, among other difficulties, there are many arbitrary parameters with regularities among them that are not explained. Of the many parameters, most are related to fermion masses and quark weak mixing matrix elements. The fermion masses have no explanation in the SM. In particular, the weak doublets of quarks and leptons of the same generation have comparable masses. Does that indicate a deep relationship between quarks and leptons and hence the strong and electroweak interactions? The existence of a GUT scale, as we discuss later in this chapter, where the interaction strengths of the strong and electroweak interactions are the same, is additional evidence for this view.

The quark and lepton weak doublets of the SM, see Fig. 1.2, are replicated three times with particles identical save for their mass. Why does this happen? Clearly we are not looking at a typical excitation spectrum, e.g. the hydrogen atom Balmer series. The dynamics must be quite unusual to have a spectroscopic series with only three terms. We also do not understand what forces are responsible for this mass splitting between generations, having exhausted the known forces (ignoring gravity) with the SM.

What is the evidence for a limited number of light generations? The primordial abundance of deuterium is related to the number of generations of neutrinos when nucleosynthesis models are used in the standard Big Bang cosmological models. The data indicate that there are three generations of light neutrinos. There is also a precision measurement available from the LEP collider. The Z decay width has been measured to high precision (see Chapter 4). The Z boson decays into quark and lepton pairs, with no flavor changing modes allowed:

$$Z \rightarrow q\overline{q}, l^+l^-, \nu\overline{\nu}. \tag{6.1}$$

The neutrinos are not detected. Measuring the "invisible" Z decay rate and dividing by the rate into neutrino pairs (see Chapter 4), we obtain the number of light neutrino species. The conclusion is that there are three and only three light species of neutrinos (below Z threshold). This finding is consistent with the one made from the prior but

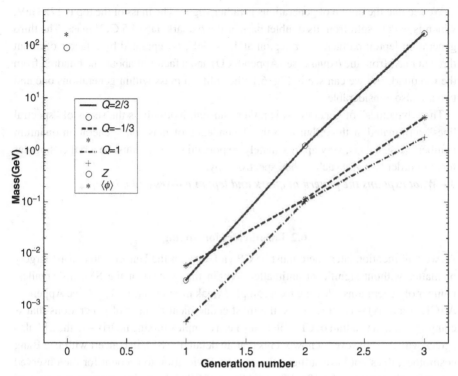

Figure 6.1 Plot of the masses of the charge 2/3 and charge −1/3 quarks and the charge 1 leptons. For this plot the neutrinos are assumed to be massless and are not shown. Also shown are the Z mass, and the Higgs vacuum field to set the electroweak scale.

weaker measurement derived from the primordial deuterium abundance:

$$N_\nu = 3. \tag{6.2}$$

These experimental facts have little or no known explanation. The dynamics, which leads to the existence of three generations, is something where we have almost no clue. Therefore, this question has no answer at present, and the paucity of hints indicates we are unlikely to find an explanation in the near future. The answer to the question put by I. I. Rabi when the muon was discovered, "who ordered that?," continues to elude us even after many years.

The quarks and charged leptons display a similar "generation" structure. There are three "generations" of quarks and leptons that have identical interactions and different masses. Note in Fig. 6.1 that there are five orders of magnitude in mass from the electron to the top quark (see Fig. 1.1 also). In fact, what we mean by a generation is simply the replication of the lowest mass electron, electron neutrino and up quark, down quark electroweak doublets of ordinary matter that recur at a higher mass. Our "ordinary" world of matter consists of bound states of the first generation quarks, held together by gluons.

We see that the notion of generations is rather vague. The mass of the top is 175 GeV, which is widely split from its doublet partner the b quark, at ~ 4.5 GeV mass. The third generation lepton partner, the τ lepton at 1.78 GeV, is separated by a factor of about three in mass from the b quark (see Appendix D) and a factor of about one hundred from the top quark. As we can see in Fig. 6.1, the splits in mass within generations one and two are also considerable.

The "dynamics" of generations is rather unusual. Not only is the series of "spectral lines" terminated at three, but also the dependence of mass on "generation quantum number" (Fig. 6.1) is, very approximately, exponential. There must be a rather singular force in order to cause such an odd spectroscopy.

4 – What explains the pattern of quark and lepton masses and mixing?

6.2 Parameters for mixing

As we will mention later, there must be CP violation for the Universe to consist largely of matter without significant antimatter. Within the context of the SM, the smallest number of generations allowing for a complex weak mixing matrix, $V_{qq'}$ – see Appendix A (CKM matrix) – is three. Thus, the most economical number of generations that is complex enough to admit of CP violation – i.e. a complex mixing matrix – in the SM also agrees with N_ν. However, it now seems that, in detail, the SM, in concert with Big Bang cosmology, does not have sufficiently strong CP violation to account for the observed baryon to photon ratio of $\sim 10^{-9}$. The condition of CP violation is necessary, but detailed calculations indicate that the SM is not sufficient.

In the strong interaction the colored quarks and gluons are flavorless. Therefore, the weak flavor quantum numbers must be produced in pairs since quark flavors are conserved in strong interactions. The flavors change in charge changing weak decays, the most familiar being beta decay, which at the quark level is, for example, $u \to d + W^+ \to d + e^+ + \nu_e$.

Over many years, experiments have been performed to determine the elements of the matrix $V_{qq'}$ characterizing the strength of the couplings in the weak decays of quarks. The matrix is completely phenomenological, since we are again ignorant of the dynamics that differentiates the weak eigenstates from the strong eigenstates. It is like knowing the $\vec{D}$ and $\vec{E}$ vectors in electromagnetism without having some fundamental understanding of the polarization of the medium. The CKM matrix, $V_{qq'}$, is shown approximately:

$$V_{qq'} \sim \begin{bmatrix} 1 & \theta_c & A\theta_c^3 p \\ -\theta_c & 1 & A\theta_c^2 \\ A\theta_c^3(1-p) & -A\theta_c^2 & 1 \end{bmatrix} \begin{bmatrix} u \\ c \\ t \end{bmatrix}, \; p = \rho + i\eta. \qquad (6.3)$$
$$\begin{bmatrix} d & s & b \end{bmatrix}$$

This matrix defines the strength of the weak decay transitions between the strong quark eigenstates. The matrix is unitary, which implies that the strength of coupling is

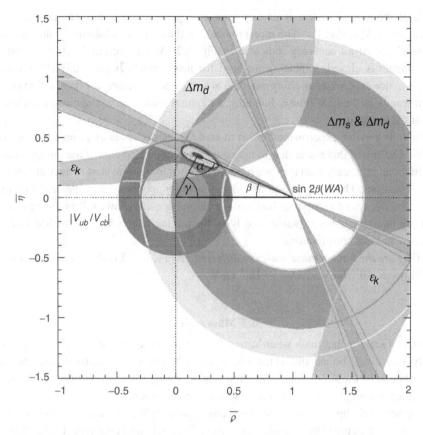

Figure 6.2 Experimental data on the elements of the CKM matrix contributing to the "unitary triangle." The elements plotted are defined to be $p = \rho + i\eta$, where p is defined in Eq. (6.3) ([1] – with permission). The shaded areas correspond to allowed regions defined by different input experimental data. The overlap of the constraints defines the unitary triangle limits.

universal, as is appropriate in a gauge theory (see Appendix A). The complex parameter $p = \rho + i\eta$ is not yet well measured (see Fig. 6.2). Numerically the parameters have values of $\theta_c \sim 0.2$ and $A \sim 1$.

A decay amplitude is proportional to the quark mixing matrix, the decay rate to the square. Clearly, the $u \to d + W^+$, $c \to s + W^+$, and $t \to b + W^+$ "diagonal" transitions are the strongest (the jargon is "Cabibbo favored"). Why is $V_{qq'}$ approximately diagonal? Why is the $b \to c + W^-$ off diagonal transition so slow, $\Gamma(q \to q') \sim V_{qq'}^2 \sim \theta_c^4$, with respect to the off diagonal transition $s \to u + W^-$, $\Gamma \sim \theta_c^2$? Is $V_{qq'}$ complex? Unitary? Does Im(p) "explain" CP violation? What is the dynamics of weak decays between generations? How can we compute the elements of $V_{qq'}$? Why is $\theta_c \sim 0.2$? There is clearly a pattern here, but we simply have no clue yet as to how to answer any of the questions that we can so easily raise.

Since the mixing matrix is unitary, each row gives a constraint. The conventional one refers to V_{ub}, V_{cb}, V_{tb}. The measurements of the complex elements of the "unitary triangle" are of an accuracy indicated in Fig. 6.2. At the present level of precision, the triangle is closed, indicating no need for new physics beyond the SM. Clearly, with a major experimental effort mounted at several accelerators, the data will improve significantly in the near future. Just now, we cannot draw a more definitive conclusion about CP violation.

There is a major experimental effort to study weak decays at electron–positron colliders. The aim of this research is to map out the complex elements of the mixing matrix much more accurately than they are presently known in order to start to answer some of these questions. The current point of attack is to determine the p parameter and therefore see if the decays of composite hadrons containing b quarks have CP violating effects that can be consistently explained solely by the mixing matrix, $V_{qq'}$, of the SM without any new physics contributions.

5 – Why are the known mass scales so different? $\Lambda_{\mathrm{QCD}} \sim 0.2\,\mathrm{GeV} < \langle\phi\rangle \sim 174\,\mathrm{GeV}$ $\ll M_{\mathrm{GUT}} \sim 10^{16}\,\mathrm{GeV} < M_{\mathrm{PL}} \sim 10^{19}\,\mathrm{GeV}$.

6.3 Mass scales

The QCD scale is that mass when strong forces become strong. It is of the same order as the meson ($q\bar{q}$ bound states) masses, as might be expected because the hadrons are states bound by the strong force. We know that the strong force gets stronger as the mass decreases, leading to complete quark and gluon confinement.

The next scale up in mass is the Higgs electroweak (EW) vacuum expectation value, which is comparable to the W and Z mass scale. The final "well established" energy scale that characterizes a "known" force is gravity, which has an energy, $U_G(r) = G_N M^2/r$, to be compared with electromagnetism, $U_{\mathrm{EM}}(r) = -e^2/r$. Gravity becomes strong when the "fine structure constant" for gravity, $G_N M^2/4\pi\hbar \sim 1$, at the "Planck mass," $M_{\mathrm{PL}} = \sqrt{4\pi\hbar/G_N}$, where G_N is Newton's universal gravitational constant. We should be aware that, since we do not have a renormalizable quantum theory of gravity, we cannot reliably extrapolate classical Newtonian gravity up to the Planck mass.

Indeed, gravity is not incorporated in the SM. Its inclusion would exhaust the known basic forces that we have observed so far. What explains the enormous "desert" – a factor 10^{17} between the electroweak scale and the Planck scale? It is, in fact, very difficult to maintain such vast difference in scales in a quantum field theory because of radiative corrections.

How, in fact, can the scales remain stable in the presence of quantum loop corrections? This is called the "hierarchy problem." A dimensional argument shows that, without some tinkering, the Higgs mass suffers an enormous shift in magnitude due to graviton (the postulated spin 2 quantum of gravity) loops, $\delta M_H^2 \sim (\alpha/\pi)(M_{\mathrm{PL}}^2)$. It is clearly necessary to explore the connection between the "low" mass scales for strong and electroweak interactions and the high mass scale characteristic of gravity. The SM is not protected against large radiative corrections feeding down from this high mass scale. Many physicists feel that this problem by itself shows that the SM is not a consistent and complete theory.

6.4 Grand unification

Only recently we found out that the weak interactions are not fundamentally weak, but had the same intrinsic strength as the electromagnetic interactions. They appear to be weak because they are confined to short distances, $\lambda_W \sim \hbar c / M_W$, by the large masses of their force carriers. Therefore, beta decays, which have energy releases ~ 1 MeV, occur with very slow reaction rates.

The electroweak unification left us with only two basic forces within the Standard Model, the strong "color" force and the electroweak "flavor" force, although the unification is not complete because the Weinberg angle is not predicted in the SM but is determined experimentally. Perhaps the strong and electroweak forces are related and hence all SM forces are unified. In that case leptons and quarks are related and there would be transitions between them. The proton would then be unstable, in clear contradiction to experiment (and our continued personal existence).

The unification mass scale, M_{GUT}, of a Grand Unified Theory (GUT) must be large enough so that the decay rate for protons, $\Gamma_p \sim 1/M_{\text{GUT}}^4$, is less than the rate limit set by experiment. There is no fundamental symmetry imposing a conservation law that we know of that requires proton stability. "Baryon conservation" is something simply put in by hand. What mass scale is there where the strong, weak, and electromagnetic forces are of equal strength? In order to answer that question we need first to explore how the strength of a force depends on mass scale.

The coupling constants "run" in quantum field theories due to vacuum fluctuations. The mathematical detail for "running" the couplings has been deferred to Appendix D. We assert here that we know how to "evolve" or "run" the coupling strengths with mass scale. We start at the Z mass. Let us see where the running of the couplings of the three forces in the SM leads us. There are three and not two because there are three distinct gauge groups, SU(2) of the weak interactions, U(1) of the electromagnetic, and SU(3) of the strong, and each gauge group has a universal coupling constant. The Weinberg angle was determined experimentally.

In general, the strength of the interaction in quantum field theories depends on the distance probed. We expect that a fine structure constant varies "generically" with mass scale Q as $1/\alpha(Q^2) = 1/\alpha(m^2) + b[\ln(Q^2/m^2)]$. A particular theory, SU(3) – strong, SU(2) – weak, U(1) – electromagnetic, defines the b parameters, which represent the effects of specific quantum loops of bosons and fermions comprising that theory and its couplings.

In electromagnetism the e^+e^- vacuum pairs shield the "bare" charge, which means that electromagnetism gets stronger at shorter distances; $b = -2n_f/12\pi$, where n_f is the number of fermions that can make virtual pairs at a scale Q. In SU(3) the strong interactions become weak at short distances. This is because the gluons themselves carry a color charge whereas the photon is uncharged. Likewise the W and Z, SU(2), self-couple having triplet vertices such as $\varphi_Z \bar{\varphi}_W \varphi_W$, $\varphi_W \bar{\varphi}_W \varphi_W$ – because they carry weak "charge." Thus we expect that the SU(2) coupling strength also gets weaker with increasing mass scale due to an anti-screening of the weak charge.

We use precision data at a mass M_Z to look for possible unification of the strong, electromagnetic, and weak forces. A representative data set is quoted in Eq. (6.4). The labels for the couplings are the SU(N) number N. The strong and electromagnetic values were already given in Chapter 4. For technical reasons of the group structure we must use 3/5 of the inverse of the electromagnetic coupling constant, $1/\alpha(M_Z) = 128.3$, minus the weak coupling. The weak coupling constant is $\alpha_W \sim 1/30 = \alpha_2$, as quoted in Appendix A.

$$\alpha_3^{-1}(M_Z) = 8.40 = 1/0.119,$$
$$\alpha_2^{-1}(M_Z) = 29.67,$$
$$\alpha_1^{-1}(M_Z) = \left(\alpha^{-1}(M_Z) - \alpha_2^{-1}(M_z)\right)(3/5) = 59.2. \tag{6.4}$$

We then "run" the constants with b values, $b_3 = (33 - 2n_f)/12\pi$, $b_2 = (22 - 2n_f - \frac{1}{2})/12\pi$, and $b_1 = -2n_f/12\pi$ (see Appendix D). The fermion loops contribute the same negative (screening) constant for all three coefficients. The strong and weak b coefficients have, in addition, anti-screening terms due to the "charged" bosons, which dominate the overall behavior. A factor of $-1/2$ in b_2 is due to a Higgs boson loop contribution. The reader is most strongly urged to use the information provided here and "run" the constants for himself or herself. The experience that is derived for the sensitivity of the couplings to large mass scales is well worth the effort.

Keeping track of the number of "active" fermions (fermions with masses less than 1/2 the mass scale Q), n_f, we arrive at the coupling constant behavior as a function of mass given in Fig. 6.3. The three forces approximately converge to a value, $\alpha_{GUT} \sim 1/43$, at a mass of $M_{GUT} \sim 10^{14}$ GeV. This is a very non-trivial result. The forces appear to be unified at a very high mass scale, one which is not terribly far from the Planck mass.

The result of following where the run leads us is a second implication that there is no new physics that intervenes observably over an enormous range in masses, from the Z mass to the GUT mass scale. That is another extremely non-trivial conclusion which follows if we accept the reality of grand unification.

6 – Why is charge quantized?

There appears to be approximate unification of the couplings at a mass scale $M_{GUT} \sim 10^{14}$ GeV. The forces that we observe to be distinct in the SM at energies <1 TeV are manifestations of the same GUT force. Since the strong force is what distinguishes quarks from leptons, that must mean that quarks and leptons are in some real sense the same particles. Therefore, we should combine quarks and leptons into GUT multiplets, where the simplest possibility for a GUT symmetry group is SU(5). In some way, with dynamics yet unknown to us, the SU(5) group breaks down into SU(3), SU(2), and U(1) subgroups at our present day mass scales.

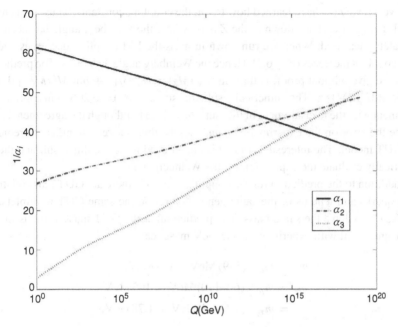

Figure 6.3 Running of the inverse of the SM coupling constants as a function of the mass scale starting at the Z mass. We run both up and down in mass from the Z roughly between limits of (Λ_{QCD}, M_{PL}).

A possible SU(5) fundamental representation for the first generation is shown below. Remember the three colors possessed by quarks, which means that the d quark appears three distinct times in the multiplet.

$$[d_R \, d_B \, d_G \, e^+ \nu_e] : 3(-1/3) + 1 + 0 = 0. \tag{6.5}$$

This seemingly innocuous statement has far reaching consequences. The sum of the projections of a group generator in a group multiplet is zero. For example, in quantum mechanics the angular momentum projection sum of m is zero for a multiplet labeled by angular momentum, $\sum_{-\ell}^{\ell} m$. Charge, Q/e (charge being the GUT coupling), must be quantized in units of the electron charge. In addition, we see that quarks must have $1/3$ fractional charge because there are three colors of quarks — SU(3). We now understand why charge must be quantized and why quarks have $1/3$ integral charge. It is because quarks and electrons are related in SU(5). In the SM we recall that charge quantization was simply put in by hand.

In addition, the unification of the three coupling constants allows us to predict the relationship between the electromagnetic and weak couplings. Recall that we simply introduced the Weinberg angle and were obliged to take its value from experiment. Now, however, we know that the GUT has a single gauge coupling constant. Thus, α and α_W must be related. The SU(5) prediction is that $\sin \theta_W = e/g_W = \sqrt{3/8}$, $\sin^2 \theta_W = 0.375$. This prediction clearly only applies at the GUT mass scale, M_{GUT}.

However, we now have learned how to run the coupling constants. Thus, we can take the GUT prediction back down to the Z mass where the Weinberg angle has been very accurately measured. When we run down in mass the EM coupling decreases and the weak coupling increases (Fig. 6.3). Hence the Weinberg angle decreases. The prediction, which we give without proof, is that $\sin^2 \theta_W(M_Z^2) \sim (3/8)/[1 + b\ln(M_Z^2/M_{GUT}^2)]$, with $b = 55\alpha(M_{GUT}^2)/18\pi$. The numerical result, that $\sin^2 \theta_W(M_Z^2) = 0.206$, is in approximate agreement with the measurement of θ_W, $\sin^2 \theta_W = 0.231$, although the agreement is well outside the error on the experimental data. Clearly, this is a very significant prediction of a GUT model. The interested student is encouraged to derive this result and then to numerically evaluate the expression for the Weinberg angle.

In addition to the prediction for the coupling constants there are GUT mass relations. Since quarks and leptons of the same generation are in the same GUT multiplets, see Eq. (6.5), they have the same mass. The prediction, at the GUT mass scale, is in only rough agreement with experiment at the GeV mass scale.

$$
\begin{aligned}
m_d &= m_e & (3\text{–}9)\,\text{MeV} &= 0.5\,\text{MeV}, \\
m_s &= m_\mu & (60\text{–}170)\,\text{MeV} &= 105\,\text{MeV}, \\
m_b &= m_\tau & (4.1\text{–}4.8)\,\text{GeV} &= 1.78\,\text{GeV}.
\end{aligned}
\tag{6.6}
$$

It is difficult to precisely define the masses of the permanently confined quarks, as they are not an observable of an asymptotically defined quantum state. Therefore, in Eq. (6.6) a range of possible masses is indicated. Still, these relations are not well satisfied. They simply validate what we mean by "generations" – a pair of quarks and a charged lepton of "similar" mass.

There is some progress that can be made by taking the prediction, which is valid at the GUT scale, and then evolving the masses down to currently available energies. This procedure leads to generally improved agreement. We note in Fig. 6.1 that typically the quarks are heavier than the leptons. That fact can be roughly understood because the quarks have strong interactions, so that the quark masses, "run" from the GUT scale to the GeV scale, evolve more rapidly than the lepton masses, rather as the coupling constants do. Therefore, the quarks are expected to be heavier than the corresponding charged leptons. For example, the successful prediction that $m_b \sim 2.9 m_\tau$ follows from SU(5) after running the masses to a scale ~ 1 GeV (see Appendix D).

However, until the GUT gauge group is experimentally known and until the assumed GUT breaking mechanism is understood, the question of quark/lepton mass relations will not yield precise predictions.

7 – Why do neutrinos have such small masses?

The neutrinos in the SM were assumed to be exactly massless, whereas we only know for sure that their masses are quite small on the scale of lepton and quark masses. This assumption is largely a question of economy, because there is no gauge condition requiring a massless neutrino. In contrast, the gluon and photon are gauge bosons and are required to be massless by the exact and unbroken gauge symmetry of SU(3) and U(1).

There is, therefore, no surprise if neutrinos possess mass and no problem absorbing a massive neutrino into the SM, just as massive quarks and charged leptons are basic particles in the SM. At worst, there are another three mass parameters and another four parameters characterizing another weak mixing matrix $V_{ll'}$. Note, however, that if neutrino mass exists there can be flavor changing leptonic reactions, just as there are for quarks. For example, $\mu \to e + \gamma$, $\mu \to e + e + e$ are then allowed reactions. At present, no such muon decay modes have been observed. However, GUT theories naturally possess lepton number and baryon number violation.

Direct kinematic measurements of neutrino masses yield results consistent with zero. There is, however, an experimental reason for imagining that a small neutrino mass might exist. The critical mass density for the Universe is ~ 5 p/m^3. Below that density the Universe will continue to expand forever. At that density the Universe is "flat." Experiment, for example the cosmic microwave background temperature anisotropy, indicates that the Universe is flat. Because the observed density of ordinary baryonic (i.e. neutrons and protons) matter is very small, we need a candidate to supply the mass needed to make the Universe flat.

The photon (approximately equal to the neutrino) to baryon ratio is known from the cosmic background blackbody radiation measurements to be $\sim 10^9$. Therefore, if neutrino masses of ~ 5 eV existed, they would supply the missing critical mass density required for a flat geometry for the Universe. We will see that the mass differences recently observed for neutrinos are much less than 5 eV, so that this explanation for the missing mass density is probably not viable.

The GUT hypothesis allows us to make a statement about why the neutrino masses might naturally be light. There are two widely separated mass scales, the QCD/EW and the GUT. Neutrinos are also unique among the fermions because, being uncharged, they can be their own antiparticles. This property makes neutrino mass creation potentially different from that for the other fermions. Assuming there are both active light neutrinos and inactive heavy neutrinos with masses comparable to the GUT scale, we state without proof that it is natural to have neutrinos with "small" characteristic masses, which means small on the quark scale. Using typical values for the masses of the three quark generations, we expect a generational hierarchy for neutrino masses.

$$m_\nu \sim m_q^2 / M_{GUT} \sim 10^{-12} - 10^{-6} - 10^{-2} \, \text{eV}. \tag{6.7}$$

There is thus an assumed natural "generation" structure for the neutrinos, which follows from the quark mass regularities gathered under the concept of generations. Recent neutrino oscillation results indicate a non-zero neutrino mass difference of ~ 0.05 eV. The neutrinos then "mix" or change "flavor" with time, much as neutral B or K mesons "mix." For a neutral particle with a fixed momentum, different neutrinos would have different energies, $E = \sqrt{P^2 + m^2} \sim P + m^2/2P$. As time goes forward after production, the state would oscillate in flavor since the states would have different frequencies, $E = \hbar\omega$, and the "beat frequency" between two states, $\Delta\omega \sim \Delta m^2 / 2P\hbar$, depends on the difference in energy between the two states of different mass. There is an extensive worldwide experimental program in place to study neutrino oscillations at present. Unfortunately,

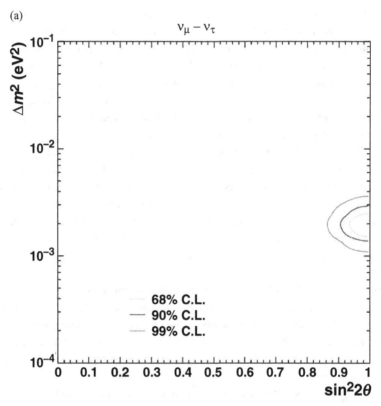

Figure 6.4 Experimental data on the mixing of neutrinos that oscillate in flavor with time. The allowed areas of mixing angle and mass difference squared are shown for different experiments and different flavor of neutrino ([2] – with permission); (a) refers to atmospheric neutrinos, while (b) refers to solar neutrinos.

the topic is beyond the scope of this text and we just indicate some of the highlights of the results of this program.

Data on neutrino oscillations are shown in Fig. 6.4. The mass differences between weak eigenstates are comparable with the estimate made in Eq. (6.7) for the third generation neutrinos. The atmospheric neutrino oscillation result is $\Delta m_{atm} \sim 0.05$ eV. The other generation neutrinos are expected to be lighter. Indeed, the solar neutrino data set indicates that a substantially smaller mass difference is responsible, $\Delta m_{sol} \sim 0.007$ eV.

We note that in the case of neutrinos the mixing appears to be nearly maximal, $\sin^2 2\theta \sim 1$, while for quarks the mixing was small and the mixing matrix was almost diagonal. We really have no clue yet as to why the quark and lepton mixings are so different.

It is not our purpose to expound on neutrino oscillations, merely to note that such oscillations require a non-zero mass for the neutrino. The GUT hypothesis explains why the masses are very small with respect to the masses of the other SM particles, and the observation of neutrino mass gives strength to the hypothesis of a large, approximately GUT,

(b)

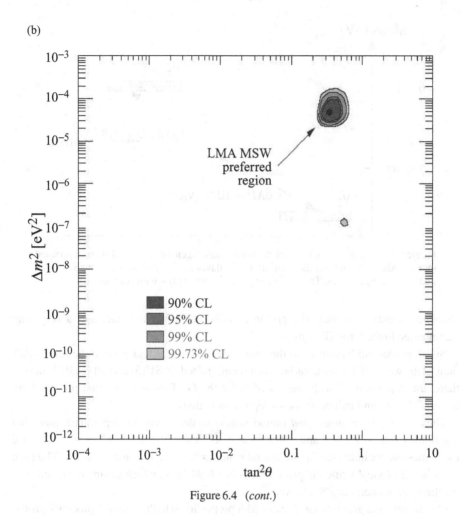

Figure 6.4 (*cont.*)

mass scale. It is not yet experimentally determined how we go from mass differences and mixing parameters to the masses of the weak eigenstates themselves. Recent precision data on the cosmic microwave background imply the limit $m_v < 0.24$ eV. One solution among many is shown in Fig. 6.5 where the masses of the weak eigenstates and the mixture of the leptonic flavors in that eigenstate are indicated.

From the cosmic microwave background we know the present neutrino temperature is $\sim 1.9°$ K and the number density is $\sim 300/\text{cm}^3$. Therefore, with the mass quoted above, the neutrinos cannot be the candidate for the "dark matter" which we discuss below because they would contribute only ~ 0.015 of the critical mass density of the Universe.

8 – Why is matter (protons) approximately stable?

There is no gauge motivated conservation law making protons stable. Baryon conservation is simply imposed in the SM by requiring, ad hoc, the absence of quark to lepton transitions. The GUT hypothesis leads us to a more incisive reason for the apparent

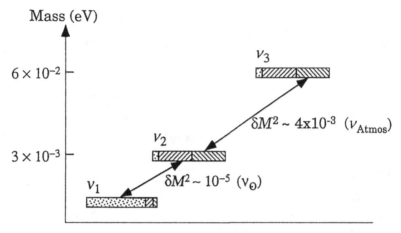

Figure 6.5   A possible scheme of neutrino weak eigenstates, v_i, and their associated masses. Also indicated are the fractions of the flavor eigenstates, v_e, v_μ, v_τ, which make up the weak eigenstates. The mixing of states is large ([3] – with permission).

absolute stability of matter. The proton is indeed unstable, but possesses a very long lifetime due to the large GUT mass.

Since quarks and leptons have the same GUT couplings and exist in the same GUT multiplets, we expect transitions between them. Indeed, in SU(5) and other GUT models there are "leptoquarks" with masses of order the GUT mass scale, that possess both flavor and color and induce quark $\leftrightarrow$ lepton transitions.

Thus we expect protons (*uud* bound states) to decay via the leptoquark mediated reactions $u + u \rightarrow e^+ + \bar{d}$ and $u + d \rightarrow v + \bar{d}$. Hence $p \rightarrow e^+ + \pi^\circ$ or $v + \pi^+$ since the π mesons are quark–antiquark bound states, $\pi^\circ = u\bar{u}$, $d\bar{d}$ and $\pi^+ = u\bar{d}$. The pion mass is $\sim$0.14 GeV while the proton mass is $\sim$0.94 GeV, which means the reaction is exothermic, or energetically allowed.

On dimensional grounds, i.e. decay width proportional to the virtual leptoquark propagator squared $\sim M_{GUT}^{-4}$, the proton lifetime should be $\Gamma_p = 1/\tau_p \sim \alpha_{GUT}^2 (m_p/M_{GUT})^4 m_p$ or $\tau_p \sim 4 \times 10^{31}$ yr. The estimate is in direct analogy to the estimate we made previously for the muon lifetime. The expected lifetime is a very long time since the age of the Universe is "only" $\sim 10^{10}$ yr. Therefore, matter is operationally, not absolutely, stable in this model.

The easiest final state to use in searching for proton decay is $e^+ + \pi^\circ$. The current experimental limit on the proton lifetime is $\sim 10^{32}$ yr. The limit is in disagreement with a much more careful estimate of the p decay lifetime in simple SU(5) GUT models. Thus we need to look a bit harder at the grand unification scheme. We have gained some insights about the open questions we had, but the unification is not actually as good as we might have hoped. We will seek improvements.

9 – Why is the Universe made of matter?

The present state of the Universe is very matter–antimatter asymmetric. Basically, there is no evidence for any primordial antimatter in the Universe. For many years we have known that the necessary conditions for such an asymmetry are that CP is violated, that baryon number is not conserved, and that the Universe went through a phase where it was out of thermal equilibrium. Now, we have already discussed the fact that the existence of three generations allows for CP violation and the initial data on the "unitarity triangle" (see Fig. 6.2) indicate that the CKM matrix has complex elements. CP violation has been observed in both K and B meson decays.

The GUT has, of necessity, baryon non-conserving reactions due to the transitions induced by the leptoquarks. We have already assumed that they are heavy, in order to explain the quasi-stability of the proton. Thus the chance to explain the matter asymmetry of the Universe exists in GUTs, although agreement of the data on the baryon to photon ratio, $N_B/N_\gamma \sim 10^{-9}$, with a detailed calculation is probably not possible. At least we have made some progress in that the dominance of matter arises naturally in a GUT model and is not simply an ad hoc assumption. Unfortunately, the SM does not contain sufficiently large CP violation.

6.5 SUSY – p stability and coupling constants

We know that there are some problems (see Fig. 6.3) with precise unification of the coupling constants and the detailed limits on the proton lifetime. These problems, and others like the Weinberg angle, can be solved by invoking a new hypothesized symmetry of Nature, called supersymmetry (SUSY). This is a symmetry that relates fermions and bosons, something that we have no indication of or hint of in the SM.

The generators of this symmetry contain both the familiar Poincaré space-time generators and a spinor connecting spin J states to J-1/2 states. Naturally, the realization of this symmetry in Nature would mean that there are super-partners of all the SM particles that differ by $\frac{1}{2}$ unit of spin. There is no experimental evidence for any of these partners, so the symmetry must be badly broken so as to give a large mass, at present experimentally inaccessible, to all the supersymmetric particles. Present limits on the mass of SUSY partners of quarks are $\sim$200 GeV. So far we have made no progress, at the expense of doubling the number of fundamental particles. Why would we embark on this daft seeming, experimentally unmotivated, enterprise?

Recall that in a quantum loop calculation the fermions and bosons contribute with opposite signs (see Chapter 4 where top increases the W mass while Higgs contributions decrease the W mass). Since each fermion now has a boson super-partner with the same mass, unbroken SUSY is very stable under radiative corrections since the loop contributions of the partners cancel.

The loop integral also depends on the mass of the particles in the loop (see Chapter 4). Therefore, "broken" SUSY will help solve the "hierarchy problem" – the radiative stability of the two widely different mass scales (EW and GUT) – only as long as the masses of the super-partners are not too large. We have traded radiative stability from the GUT mass scale for a proliferation of new and unobserved particles. We argue that the masses of broken SUSY must appear in the mass range $\sim$(100, 1000) GeV if SUSY is to solve

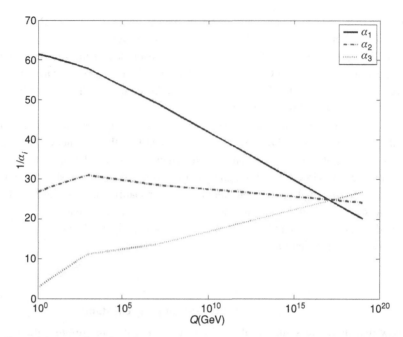

Figure 6.6   Running of the inverse of the SM coupling constants as a function of the mass scale with a partner SUSY spectrum added at a mass of 1 TeV.

the hierarchy problem. This mass range is accessible at the LHC, so that SUSY will be very actively looked for in LHC experiments.

This is fine, but is there presently any "evidence" for a "SUSY–GUT"? Alas, we have only rather indirect indications. Let us return to the issue of grand unification. We add SUSY particles to the spectrum and look again at the running of the couplings. The detailed running behavior is altered by these new particles in the loops. The evidence for unification is now stronger, with $M_{\mathrm{GUT}} = 2 \times 10^{16}$ GeV and $1/\alpha_{\mathrm{GUT}} \sim 24$. The graphical representation of the situation is shown in Fig. 6.6. Note the "kink" in the behavior of the running couplings when the SUSY partners become "active" at ~ 1 TeV in mass. The low mass behavior is identical to that shown for the SM in Fig. 6.2.

Clearly, the case for unification is much improved over the SM results. In particular, the increased GUT mass, and the strong dependence of the proton lifetime on that mass, $1/M_{\mathrm{GUT}}^4$, solve the problem we had with the experimental limit on the proton lifetime. Given the logarithmic dependence of the couplings, it is difficult to impossible to conclude that the SUSY "kinks" occur at a particular mass. In the case displayed, 1 TeV is used, but there is little sensitivity to that mass within wide limits.

The prediction for $\sin^2 \theta_{\mathrm{W}}$ at the Z mass is also altered because the evolution down from the gauge coupling value of 3/8 at the GUT mass scale is changed. The starting point GUT mass has been increased and there are now additional SUSY particles in the loops for $Q > 1$ TeV. The prediction goes from 0.206 to 0.23, significantly improving the agreement with experiment, which obtains the value of 0.231. This agreement of experiment and theory for the measured Weinberg angle to a few percent with SUSY is strong indirect evidence for SUSY.

The hierarchy problem has to do with the existence of two mass scales that are radically different. It is difficult to maintain the lower mass scale, say for the Higgs boson mass, in the presence of radiative corrections for loops containing particles with the higher mass scale. We have already mentioned this in the context of the Planck mass. Now we hypothesize that there is a somewhat lower GUT mass scale, which intervenes and basically erases information about the Planck scale if we are at lower masses than the GUT scale. The loop corrections to the Higgs mass are quadratically divergent. Going from GUT mass M_{GUT} to the electroweak scale, the Higgs mass shift is huge:

$$\delta M_H^2 \sim (\alpha_{GUT}/\pi) \left(M_{GUT}^2\right). \tag{6.8}$$

To maintain the Higgs mass in the absence of SUSY two numbers of order M_{GUT} must subtract to yield a small number, M_H, which is very "fine tuning." In a SUSY GUT, since equal mass bosons and fermions contribute to these loop integrals with opposite signs, the large radiative corrections are canceled to very high order. Thus SUSY solves the "hierarchy problem." With SUSY masses at a much lower mass scale, the Higgs mass gets radiative corrections due to the differences of the masses of the SUSY, M_{SUSY}, and SM, M, partners:

$$\delta M_H^2 \sim (\alpha_{GUT}/\pi) \left(M_{SUSY}^2 - M^2\right). \tag{6.9}$$

There are two predictions that are very relevant for LHC experimentation. First, SUSY only solves the hierarchy problem if M_{SUSY} is < 1 TeV, and hence these states will most likely be accessible at the LHC.

Second, we assert without proof that some SUSY models constrain the Higgs mass to be $M_H < M_Z$. Radiative loop corrections then imply that the Higgs mass is increased from the Z mass by top and SUSY top partner particles in the loop, Eq. (6.10). An upper limit, $M_H < 130$ GeV, is then approximately derived, which is somewhat more stringent than the limit we have already quoted in Chapter 4.

$$\delta M_H^2 \sim 3\alpha_W/2\pi (m_t/M_W)^2 m_t^2 \left[\ln \left(M_{SUSY}^2/m_t^2\right)\right]. \tag{6.10}$$

Thus if SUSY is true, a light Higgs is expected in some simple models, nay required, which is very accessible at the LHC (Chapter 5). This prediction of SUSY will be verifiable in the very near future. In addition, we know that the parameter λ in the Higgs potential sets the Higgs mass and "runs" (see Chapter 5) and that it must be positive for there to be a non-zero value of the vacuum field (vacuum expectation value). We assert without proof that a heavy top quark mass is needed in SUSY models if the Higgs mechanism is to be preserved. The observed large top mass (Fig. 6.1) can be seen as another successful prediction of SUSY. It is also true that large CP violations occur naturally in SUSY models since many of the new coupling constants introduced by SUSY can be complex. Therefore, SUSY would improve the deficiency of CP violation strength that is present in the SM.

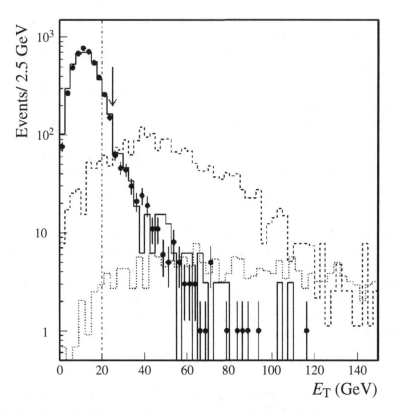

Figure 6.7   Missing transverse energy distribution for events with a photon and at least two jets in the final state in D0 data. Also shown are the signals expected for SUSY quarks of 150 (dashed) and 300 GeV (dotted – cross section times 10) ([4] – with permission).

6.6 SUSY – cross sections at the LHC

SUSY particles have already been carefully searched for at the Tevatron and we could have introduced them in our discussion of Tevatron physics in Chapter 4. We do so now, as this fits the flow of the narrative. Normally it is assumed that there is a quantum number associated with SUSY, which, like flavor, requires pairs of particles to be produced in the interactions of SM particles. Unlike flavor, the symmetry is assumed to be exact, so that the lightest SUSY particle (LSP) is absolutely stable. Therefore, assuming the LSP is neutral and weakly interacting, most SUSY searches use jets (from cascade decays down to the LSP) and missing transverse energy (taken off by the LSP) in setting limits on SUSY particle masses. In what follows, supersymmetric partners of the known particle spectrum are indicated by a tilde. Thus $\tilde{q}$ indicates a supersymmetric quark or a "squark." There is no evidence yet at the Tevatron collider for a SUSY signal. A typical spectrum is shown in Fig. 6.7.

The "background" from SM processes falls off rapidly (it is largely missing energy due to the mis-measurement of the jet energies or $W \to \nu_e + e$, $Z \to \nu_\ell + \bar{\nu}_\ell$ decays),

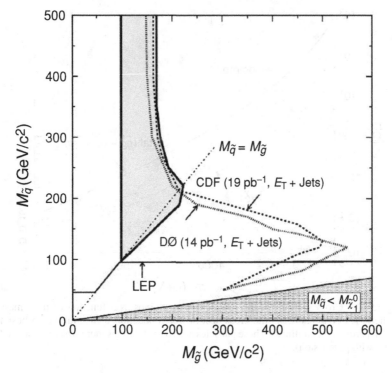

Figure 6.8   Excluded contours at the 95% confidence level for SUSY partners of the quarks (squarks) and gluons (gluinos) from Tevatron and CERN collider experiments in a minimal SUGRA SUSY model ([5] – with permission). The shaded exclusion region uses CDF data on like-sign dileptons plus jets and missing energy.

leaving the spectrum at large missing energy dominated by possible signals from the SUSY partners of the quarks. Clearly, SUSY quarks of 150 GeV mass are excluded, while 300 GeV mass is not totally excluded by this data set. Higher statistics data from the upgraded Tevatron will push out the mass limits. The present limits on SUSY masses, in the context of a particular SUSY model chosen from a plethora of possible models, are shown in Fig. 6.8.

Clearly, masses of ~200 GeV and below are excluded. Since we argued that SUSY particles must have masses less than 1000 GeV if they are to solve the hierarchy problem, this level of exclusion is already very significant. Unfortunately, the 1 TeV upper limit is not very crisp (see Fig. 6.6), so that we should be prepared, at the LHC, to search well above it if we are to definitively exclude SUSY as a hypothesis that solves the hierarchy problem.

Let us imagine how to continue this search at the LHC. The cross sections for squarks and gluinos (SUSY partners of quarks and gluons) are large because they have strong couplings. The couplings of the SUSY particles are the same as those of their SM partners except for the kinematic effects of mass. The equality of the forces is needed for loop cancellations, and is intrinsic to SUSY. Dimensionally, the cross section for strong

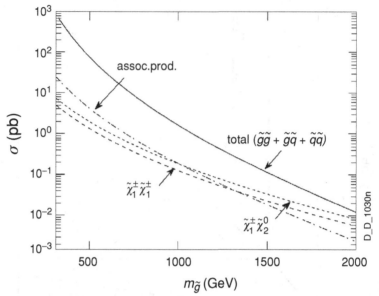

Figure 6.9   Cross section for the production of gluinos as a function of their mass
at the LHC. Also shown is the cross section for production of neutral gauge boson
SUSY partners (neutralinos). These particles are 10 to 100 times more weakly produced
([6] – with permission).

production of a pair of mass M particles is $\hat{\sigma} \sim \alpha_s^2/(2M)^2$ or ~ 1 pb for $M = 1$ TeV.
This level of cross section is quite observable at the high luminosity available at LHC
experiments (100 000/pb yr.), and ~ 0.1 pb (α_w/α_s) (1 pb) for neutralino.

A complete calculation of the cross section as a function of SUSY mass is shown in
Fig. 6.9. The cross section for SUSY quarks and gluons is, indeed, approximately 1 pb,
for a 1 TeV mass.

For a 500 GeV SUSY gluino, the cross section is 100 pb. Thus, running only a month
at 1% of the design luminosity, 10 000 SUSY gluino pairs are created. Clearly, searching
for strongly produced SUSY particles will be a major part of the very early LHC physics
program. The experimenters must be prepared for incisive searches as soon as the LHC
begins to function.

6.7 SUSY signatures and spectroscopy

We know that the cross section, at least for strongly interacting SUSY partners, is large
enough for discovery at the LHC. The question is, what are the signatures for triggering
the apparatus (see Chapter 2) on SUSY particle production? For squarks and gluinos a
straightforward method is to look at jets and missing energy. A possible set of decay
modes is shown in Fig. 6.10. Multi-jets, leptons, and missing energy in coincidence
supply a rather spectacular and unique signature on which to trigger and then search for
signals.

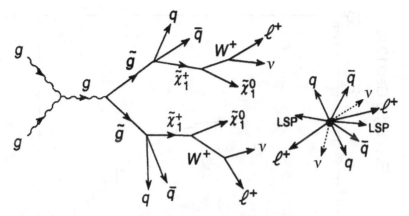

Figure 6.10   Schematic representation of gluino pair production and sequential decays. The end of the decay chain comes with the emission of two LSP neutralinos. The cascade decays result in a final state with four quark jets + two leptons (same sign) + missing transverse energy from the LSP, $\tilde{x}_1^0$, and the neutrinos ([7] – with permission).

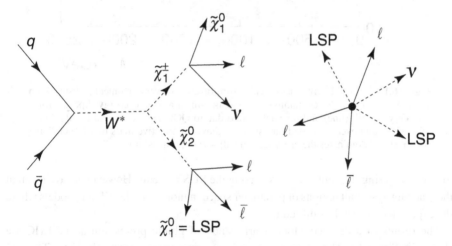

Figure 6.11   Schematic representation of Drell–Yan production of a highly off-mass shell, W, W^*, which virtually decays into a gaugino pair. The subsequent decay into leptons, neutrinos, and LSP leads to a final state with no jets, missing transverse energy, and three leptons, which is a very clean SUSY signature ([7] – with permission).

For SUSY gauge partners there are also very unique signatures. A schematic representation is shown in Fig. 6.11. The Drell–Yan production of an off-mass shell W results in the decay to a gaugino pair. The subsequent cascade decays to the LSP neutral gaugino then results in no jets (we can even veto on jets in the trigger if needed), three leptons, and missing transverse energy.

Fundamentally, the SUSY searches are fairly straightforward at the LHC. There are, of course, other decay modes and other signatures. However, as long as pair production of SUSY particles is assumed, then the existence of a LSP makes missing energy a powerful

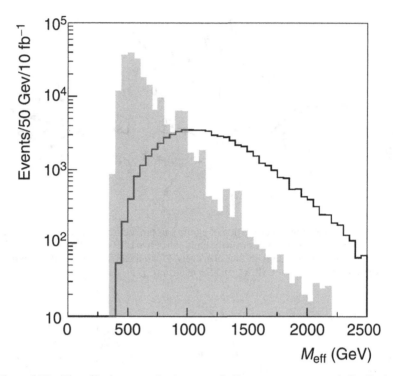

Figure 6.12 The effective mass (scalar sum of all transverse momenta) distribution for events at the LHC containing jets and missing transverse energy. Spectra for a 600 GeV squark (line) and for background due to QCD jet events, and for W/Z plus jet background events (shaded) are shown. Above an effective mass of 1000 GeV the SUSY signal dominates the cross section ([8] – with permission).

tool for triggering the detector and selecting the SUSY events. However, a caveat is that there are always small regions of parameter space in more complex SUSY models where the experimental search is difficult.

The results of a detailed Monte Carlo study of gluino production at the LHC are shown in Fig. 6.12. The trigger is on missing transverse energy plus jets. The SM "background" from QCD jet production with a missing transverse energy caused by mis-measurement of the jet energies (see Chapter 2) falls rapidly with transverse energy. Processes with $W/Z +$ jets, for example top pairs, contain real missing E_T, but occur at a lower cross section than the QCD production of jets with a subsequent experimental mis-measurement inducing a missing E_T. Clearly, above an effective mass E_T of ~ 1000 GeV, the signal from a 600 GeV gluino dominates over all the SM backgrounds (note that Fig. 6.12 has a logarithmic vertical axis).

We have checked that we can search for masses above 250 GeV until we run out of events due to the falling of the SUSY cross sections with mass. The 250 GeV mass scale was considered here since it roughly corresponds to the current SUSY mass reach of Tevatron collider experiments (see Fig. (6.8)), and at the LHC the experimenters want to pick up the search with no mass range remaining inaccessible. The LHC experiments

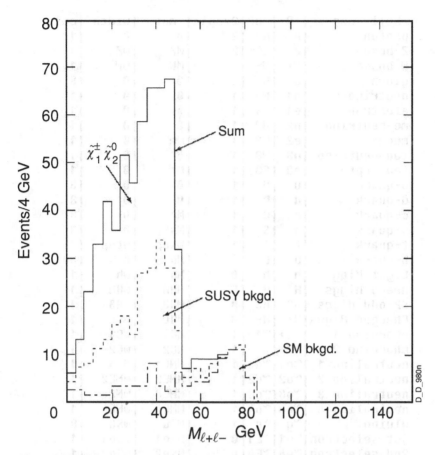

Figure 6.13 Dilepton mass spectrum for events with multiple leptons and no jets. Masses above 81 GeV are deleted. The luminosity corresponds to one year at 10% of design luminosity ($10\,\text{fb}^{-1}$). The sharp edge corresponds to the mass difference between the two neutral charginos ([9] – with permission).

can therefore pick up the SUSY search seamlessly from the CDF and D0 experiments and carry it up to ∼2 TeV in mass. The upper limit is important, because if SUSY is too heavy then it is not the solution to the "hierarchy problem."

In SUSY, there is a series of cascade decays down to the LSP. That decay topology allows us to determine some of the mass differences of SUSY particles at the LHC. In particular, there are spectacularly sharp spectral edges in specific cases. This gives us another handle on the spectroscopy of SUSY particles. An example is shown in Fig. 6.13. The distribution of dilepton masses is shown, where the events were selected to have leptons above a cut on transverse momentum and all jets were vetoed on. As one can see, there is expected to be a sharp kinematic edge corresponding to the neutral chargino mass difference (see also Fig. 6.11) with $\tilde{\chi}_2^o \rightarrow \tilde{\chi}_1^o + \ell^+ + \ell^-$. Therefore, we can go beyond the mere discovery of SUSY and learn something about the complex spectroscopy that would become experimentally available should SUSY be realized in Nature.

Full name	P	aP	2*spin	mass	width	color
photon	A	A	2	0	0	1
Z boson	Z	Z	2	MZ	wZ	1
W boson	W+	W-	2	MW	wW	1
gluon	G	G	2	0	0	8
neutrino	n1	N1	1	0	0	1
electron	e1	E1	1	0	0	1
mu-neutrino	n2	N2	1	0	0	1
muon	e2	E2	1	Mm	0	1
tau-neutrino	n3	N3	1	0	0	1
tau-lepton	e3	E3	1	Mt	0	1
u-quark	u	U	1	0	0	3
d-quark	d	D	1	0	0	3
c-quark	c	C	1	Mc	0	3
s-quark	s	S	1	Ms	0	3
t-quark	t	T	1	Mtop	wtop	3
b-quark	b	B	1	Mb	0	3
Light Higgs	h	h	0	Mh	wh	1
Heavy higgs	H	H	0	MHH	wHh	1
CP-odd Higgs	H3	H3	0	MH3	wH3	1
Charged Higgs	H+	H-	0	MHc	wHc	1
chargino 1	~1+	~1-	1	MC1	wC1	1
chargino 2	~2+	~2-	1	MC2	wC2	1
neutralino 1	~o1	~o1	1	MNE1	0	1
neutralino 2	~o2	~o2	1	MNE2	wNE2	1
neutralino 3	~o3	~o3	1	MNE3	wNE3	1
neutralino 4	~o4	~o4	1	MNE4	wNE4	1
gluino	~g	~g	1	MSG	wSG	8
1st selectron	~e1	~E1	0	MSe1	wSe1	1
2nd selectron	~e4	~E4	0	MSe2	wSe2	1
1st smuon	~e2	~E2	0	MSmu1	wSmu1	1
2nd smuon	~e5	~E5	0	MSmu2	wSmu2	1
1st stau	~e3	~E3	0	MStau1	wStau1	1
2nd stau	~e6	~E6	0	MStau2	wStau2	1
e-sneutrino	~n1	~N1	0	MSne	wSne	1
m-sneutrino	~n2	~N2	0	MSnmu	wSnmu	1
t-sneutrino	~n3	~N3	0	MSntau	wSntau	1
u-squark 1	~u1	~U1	0	MSu1	wSu1	3
u-squark 2	~u2	~U2	0	MSu2	wSu2	3
d-squark 1	~d1	~D1	0	MSd1	wSd1	3
d-squark 2	~d2	~D2	0	MSd2	wSd2	3

Figure 6.14   COMPHEP particle table (truncated for SUSY quarks at the first generation) showing the particle content of the MMSM model.

The COMPHEP program has a SUSY model available to use in evaluating cross sections. For example, in the "MMSM," the process $g + g \rightarrow \tilde{g} + \tilde{g}$ with SUSY gluons of 200 GeV mass has a LHC cross section of 3.4 nb, consistent with the cross sections shown in Fig. 6.9.

There are many complications in using the SUSY model in COMPHEP, as the number of particles is rather large. Nevertheless, SUSY decay branching fractions and SUSY production cross sections can be studied as desired.

As an example, the particle content in the MMSM (minimal) SUSY model is shown in Fig. 6.14. This rather long table has been truncated at the first generation of quarks. In addition to the SM particles, which are the first entries, there are now four Higgs particles because the Higgs "sector" proliferates in SUSY. There are also two charginos, four neutralinos, and one gluino. That completes the list of SUSY partners of the gauge bosons. The remaining entries are the SUSY partners of the SM leptons and quarks. Given the added complexity of the particle content, we do not typically invoke SUSY models in this text. The interested student can, however, profitably spend some time looking at the implications of SUSY dynamics using the additional tools provided by COMPHEP.

10 – What is "dark matter" made of? What is "dark energy"?

First we need to explain what we mean by "dark matter." The Universe appears to have a critical (or closure) energy density. The energy density of the Universe defines the geometry in general relativity, whether it has positive curvature, is flat, or has negative curvature. There are many reasons, both theoretical and now experimental, in cosmology to favor a flat solution, e.g. "inflation," and thus a "critical" energy density, which defines the transition from a closed (positive) to an open (negative) geometry.

We can try to identify this energy density with the matter that we can see. If we simply count stars, there is only $\sim$0.01 of the closure density that we can account for. Yet the Universe appears experimentally to be approximately flat (supernovae as "standard candles," and a roughly linear velocity (Doppler shift)–distance (observed brightness) relationship, for example). What is that matter made of? Parenthetically, it may seem odd, but we have no idea what form most of the energy in the Universe takes. This is a humbling statement as we begin the twenty-first century.

Instead of counting visible mass, we can try to measure the mass of an object dynamically by using Newtonian mechanics. This method has the advantage that it measures non-luminous matter too. When we try to measure the mass of a galaxy dynamically, we want to look at the orbital velocity (measured by using the Doppler shift), v, as a function of radius. Newtonian energy conservation tells us that $GM(r)/r = v^2$, where $M(r)$ is the mass found within a radius r. If we have a uniform central mass density, $M(r) \sim r^3$ and $v \sim r$. Beyond the central luminous region, if all the mass is distributed as is the luminous mass, then $M(r) \sim$ constant, and the falloff of velocity with distance is expected to be $v \sim 1/\sqrt{r}$. This situation is familiar from our own solar system and is embodied in Kepler's Laws. The square of the orbital period is proportional to the cube of the orbit radius.

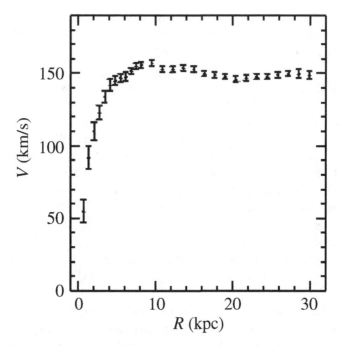

Figure 6.15   Orbital velocity of matter within a galaxy as a function of the radius from the galactic center. The velocity is observed to be constant out well beyond the luminous core of the galaxies ([10] – with permission).

Some data on $v(r)$ as a function of r for a typical galaxy are shown in Fig. 6.15. In fact, we do observe the expected linear rise of $v(r)$ with r at small values of r. However, no falloff is observed in velocity out to a radius of ~60 kpc, well beyond the luminous region of typical galaxies. Rather we see $v(r) \sim$ constant, which indicates $M(r) \sim r$ for the "dark matter," or non-luminous, contribution to galactic dynamics.

Is this evidence for the SUSY partners – the stable LSP relics of the Big Bang? The SM does not contain a candidate particle for the "dark matter," and the newly discovered neutrino mass differences that are seen in the oscillation experiments (<0.05 eV), if they are indicative of the masses themselves, are too small to reach the critical density which requires a mass of ~20 eV for neutrinos. SUSY, on the other hand, certainly provides a dark matter candidate.

In fact, the fairly heavy SUSY particles, the LSP neutralinos, also have weak cross sections, a fact that is needed to solve the "dark matter" problem. The argument goes as follows. Dark matter exists at about one third the closure density of the Universe. The neutralino "decouples" from other particles participating in the cosmic expansion when the annihilation rate of neutralinos falls below the cosmic expansion rate. Annihilation cross sections for weakly interacting particles of mass M are generically $\sigma_A \sim \alpha_W^2 / M^2$.

Thus the "relic" LSP abundance of neutralinos left over from the "Big Bang" depends on the neutralino mass, M. A larger cross section means a longer coupling time, which means, in turn, less present abundance. A limit on the LSP density at the critical density places a "cosmologically interesting" mass limit, as shown in Fig. 6.16. Numerically, it is a strong clue that a particle of mass $\sim$1 TeV must have a weak interaction cross section if it is to be the source of dark matter. SUSY therefore "naturally" has the weakly interacting neutralino as a dark matter candidate.

Experiments at the LHC can quickly set limits, in a particular model incarnation of SUSY called SUGRA, on SUSY particles such that $<$2 TeV is excluded, as seen in Fig. 6.16. Therefore, at the LHC we can probably either discover SUSY or decisively remove it as a model put forward to solve the hierarchy problem. LHC experiments will also quickly set limits on the LSP mass that span the cosmologically interesting range for dark matter.

Recently, evidence has been given that the energy density of the Universe is dominated ($\sim$70 % as of today) by "dark energy." This "stuff" has negative pressure, as does a cosmological constant, and accelerates the expansion of the Universe. There appears to be a cosmological constant that is not zero, as had been assumed by Einstein. It is fair to say that, if the evidence holds up, we do not have a clue what the "stuff" is – an exciting puzzle to fathom.

11 – Why is the cosmological constant so small?

6.8 Cosmological constants (and SUSY?)

The vacuum expectation value of the Higgs field is 174 GeV, corresponding to a mass density (a proton has 0.94 GeV mass) of $\sim$174 GeV/(0.001 15 fm)^3 $\sim$ 130 p/(0.001 fm)^3 $\sim$ 1.3 $\times$ 10^{56} p/m^3. This vacuum field appears to exist, in that the W and Z masses have been observed and precisely measured. On the other hand, the vacuum energy density of the Universe ("dark energy") is known to be near the critical value of $\sim$5 p/m^3. The electroweak vacuum expectation value of the Higgs field is therefore $\sim$10^{56} times larger, which presents us with a monumental mismatch.

Recent observations, e.g. the supernovae measurements of velocity versus distance that deviate from Hubble's linear law and indicate cosmological acceleration, support a non-zero cosmological constant with a magnitude near that of the critical density. That in itself is enormously interesting because it indicates that a vacuum energy density, such as is needed for inflation, indeed exists and is extremely small on the scale of the SM vacuum energy density.

This fact does not address the enormous disparity in the two values of the vacuum energy. We assert that a vacuum virtual loop will make different signed contributions to the vacuum energy for fermions and bosons as it does with other loops. If the couplings are SUSY related, the contribution to the cosmological constant might be reduced. Still the discrepancy is "astronomical," and we truly cannot now make any plausible scenario wherein the vacuum energy can be made to agree with experiment in the SM + SUSY GUT context.

However, if SUSY is invoked as a local symmetry, as are the other gauge symmetries in the SM, then many interesting conclusions ensue. Local SUSY theories, which are

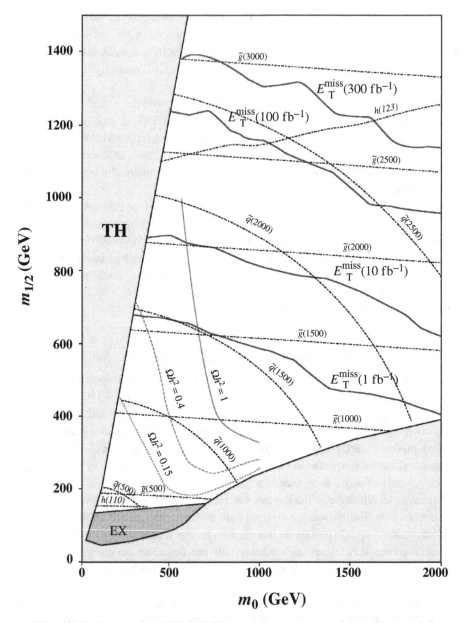

Figure 6.16   Contours in SUSY–SUGRA mass parameter space that can be excluded by LHC experiments. Existing limits are shown shaded. Contour limits for 1, 10, 100, and 300 fb^{-1} are shown (solid lines). Contours of constant squark and gluino mass are also indicated (dashed lines). In 1 year of running at full LHC luminosity, squarks and gluinos (see Fig. 6.11) will be excluded if they have masses <2 TeV. In this model, that search sensitivity then easily excludes a LSP that would be a dark matter candidate. Such candidates are well inside the $\Omega h^2 = 1$ contour ([6] – with permission).

called generically "supergravity," have both positive and negative contributions to the vacuum energy. That, in turn, means that perhaps we can have a cosmological constant consistent with observations. However, we are very, very far away from being able to make any real calculation.

12 – How does gravity fit in with the strong, electromagnetic, and weak forces?

6.9 SUSY and gravity

Since SUSY is an attractive theory, which solves the hierarchy problem, the proton decay limit, improves coupling constant unification, improves the prediction for the Weinberg angle, and supplies a dark matter candidate, it seems natural to try to make SUSY a local symmetry by analogy to the known SM gauge symmetries. A local SUSY theory, since SUSY has both spin and Poincaré generators, will be a theory of general coordinate transformations. Therefore, a local SUSY theory, in the classical limit, contains General Relativity very naturally. As we have so far been unable to incorporate gravity into the SM, this fact is of extraordinary interest. Note, however, that the theory is classical; it is not a renormalizable quantum theory.

A Planck scale and a SUSY breaking scale, $M_s \sim 10^{11}$ GeV, can be invented, which interact similarly to the neutrino "seesaw" to give masses to the SUSY partners of the SM with $\sim M_S^2/M_{PL} \sim 1000$ GeV. However, a local SUSY model of point particles, although it contains classical gravity, still is not a renormalizable field theory.

Why can't we incorporate gravity? Let us look at the "running" of the gravitational coupling constant. We make the most naïve extrapolation of Newtonian classical gravity to assign a fine structure constant for gravity. Because gravity alters the very fabric of space-time, we cannot expect such an extrapolation to the regime of strong gravity to be valid, only indicative.

Recall that the Planck scale occurs when the gravitational fine structure constant becomes strong, $\alpha_G = G_N M^2/4\pi\hbar c \sim 1$, at a mass scale $M_{PL} = \sqrt{4\pi\hbar c/G_N} = 1.2 \times 10^{19}$ GeV. For amusement we compare the "running" of the renormalizable gauge theories of the SM to this naïve extrapolation in Fig. 6.17. The quadratic energy dependence of gravity on mass is much stronger than the logarithmic variation of the SM forces. This bad high energy behavior of gravity is what makes it not renormalizable.

It is clear that there is a weak indication that the high mass scale of SM unification (the GUT scale) is not too distant from intersecting the naïve running of the gravitational force. Considering that we do not have a complete quantum theory of gravity, this fact is provocative. Perhaps with a correct quantum theory of gravity a complete unification of all the known forces is possible. Indeed, in a "string theory" – or candidate quantum theory of gravity – the appropriate scale is less than the Planck scale, thus reducing the discrepancy.

A renormalizable theory of gravity appears to be impossible with point particles. Using particles extended in one dimension ("strings") as the fundamental entities, a well-behaved theory of gravity is possible, but only in a space of high dimensionality. Not only does gravity appear naturally in this string formulation, but SUSY does also.

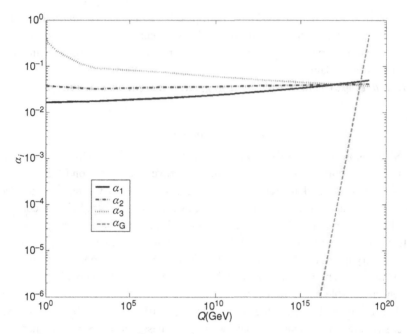

Figure 6.17   The running of the coupling constants for the SM forces having logarithmic mass dependence and gravity, thought classically to go as the square of the mass scale. There is a very approximate unification at a very high mass scale.

These other dimensions are usually assumed to be "compactified" at length scales of order the Planck length, so that we are unaware of their existence. In fact, the Standard Model gauge symmetries appear to arise, almost naturally, in some string theories with a compact subspace. More speculatively, the number of generations might be related to the topology of the compact space.

Recently, the alternative possibility of "large" extra dimensions has been raised as another solution to the hierarchy problem. These extra dimensions may also throw light on the unification of the gravitational coupling and the Standard Model gauge couplings. If gravity exists in all the extra dimensions while all the other SM gauge forces are confined to four dimensional space-time, then the scale where gravity becomes strong might be the electroweak scale of ~ 1 TeV rather than the Planck mass scale.

Gravity is known to have a $1/r$ potential only for distances greater than about 1 mm. In low energy laboratory experiments, the Newtonian potential is altered by a factor $\sim 1 + d/r^2$, where d may be such that the deviation is measurable. Laboratory experiments are now in train to study deviations of gravity from an inverse square law at millimeter length scales.

Gravity with large extra dimensions is thought to be weak because it "leaks" into the other dimensions while the SM forces do not. This is natural because string theory is only well behaved in spaces with a large number of dimensions. If large extra dimensions were the solution of the hierarchy problem, then of necessity, there should be effects of graviton exchange, which might be accessible at the new generation of colliders that probe the electroweak mass scale. One obvious signature is missing energy caused by a graviton escaping into the extra dimensions. These new phenomena will be searched for at the LHC and elsewhere. The key signatures will be reactions, like gravity, with spin 2 Lorentz character (e.g. angular distributions) and which couple to mass/energy without regard to other variables (gravity is both flavorblind and colorblind).

These "theories of everything" are, so far, almost devoid of testable predictions and are perhaps in the province of philosophy or metaphysics and not physics. Time will tell.

Summary for hadron collider physics

- The LHC will explore the full (100–1000 GeV) allowed region of Higgs masses. Precision data indicate that the Higgs is light. If the Higgs is, in fact, light then its couplings can probably be explored by observing decays into $b\bar{b}$, $\gamma\gamma$, $\tau^+\tau^-$, W^*W, and Z^*Z.
- There appears to be a GUT scale that indicates new dynamics. The GUT explains charge quantization, predicts the rough value of θ_W, allows for the matter dominance of the Universe, and explains the small values of the neutrino masses. However, it fails in p decay, precise Weinberg angle prediction, and quadratic radiative corrections to Higgs mass scales – the hierarchy problem.
- Preserving the scales (hierarchy problem) can be accomplished in SUSY. SUSY raises the GUT scale, making the p quasi-stable. The Weinberg angle SUSY prediction is in accord with the precision data. The SUSY LSP provides a natural candidate to explain the observation of galactic "dark matter." A local SUSY GUT can incorporate gravity. It can also reduce the cosmological constant problem. A common GUT coupling and preservation of loop cancellations requires SUSY mass <1 TeV. The LHC will fully explore this SUSY mass range, either definitively proving or disproving this attractive hypothesis.
- If there are extra dimensions, then the LHC is well positioned to study the TeV mass scale, where their effects should appear if they are the solution of the hierarchy problem.
- The generational regularities in mass, quark, and neutrino mixing matrix elements will probably not be informed by data taken at the LHC. We still haven't a clue "who ordered that."

Exercises

1. Combine the decay width scaling as M^5 and as $V^2_{qq'}$ to estimate the decay width of $c \rightarrow s$ with respect to that of $b \rightarrow c$. Are they comparable?
2. Evaluate the shift in $1/\alpha_R$ (Q^2/m^2) (Appendix D) from $Q = 1$ m to $Q = 1000$ m.
3. Evaluate $\alpha_1^{-1}, = 59.2$ (M_Z), at the GUT mass.
4. Evaluate $\alpha_3^{-1}, = 8.40$ (M_Z), at the GUT mass. Is it close to the coupling constant evaluated in Exercise 3?
5. Evaluate the Weinberg angle going from $\sin^2 \theta_W = 0.375$ at the GUT scale to the Z mass scale.

6. Suppose that the neutron is a bound state of *udd* quarks. Show that the fundamental decay modes $d + d \rightarrow e^- + \bar{d}, u + d \rightarrow \nu + \bar{d}$ conserve electric charge and lead to the observable decays $n \rightarrow e^- + \pi^+, n \rightarrow \nu + \pi^0$.

7. Explicitly work out the estimate for the proton lifetime for a GUT mass of 10^{14} GeV. How does it change if the GUT scale goes to 10^{16} GeV?

8. Assume that SUSY particles have the same coupling as their SM partners. Evaluate the point like cross section for a SUSY mass of 2 TeV, $\hat{\sigma} \sim \alpha_S^2/(2M)^2$ and compare with the Monte Carlo model. Does the gluon source factor, $(1 - M/\sqrt{s})^{12}$, improve the agreement?

9. Make a complete calculation of the gravitational problem of orbits around a distributed mass. Show that the velocity inside a uniform distribution goes as r, while the velocity outside the distribution goes as $1/\sqrt{r}$.

10. Show explicitly that the vacuum expectation value of the Higgs field contributes an energy density $\sim 10^{56}$ times the closure density of the Universe.

11. Show that the closure density of the Universe, if ascribed to a vacuum field, has a vacuum expectation value $\langle \phi \rangle \sim 0.001$ eV.

12. Use COMPHEP to evaluate the Z decay width and branching fraction $(Z \rightarrow 2^*x)$. Compare with the data shown in Chapter 4. What are the neutrino branching fractions?

13. Look in COMPHEP at the Standard Model parameters and find the quark mixing matrix elements.

References

1. Pitts, K., Fermilab Conf-00-347-E and hep-ex/0102010 (2001).
2. Scholberg, K., arXiv:hep-ex/0011027 (2000).
3. Fisher, P., Kayser, B., and McFarland, K., *Ann. Rev. Nucl. Part. Sci.*, **49**, 481 (1999).
4. D0 Collaboration, *Phys. Rev. Lett.*, **82**, 29 (1999).
5. CDF Collaboration, *Phys. Rev. Lett.*, **87**, 251803-5 (2001).
6. Pauss, F. and Dittmar, M., ETHZ-IPP PR-98-09, hep-ex/9901018 (1999).
7. Pauss, F., CMS Note, 1998-097 (1998).
8. Gaines, I., Green, D., Kunori, S., Lammel, S., Marriffino, J., Womersley, J., and Wu, W., Fermilab-FN-642, CMS-TN/96-058 (1996).
9. Abdullin, S., Antunovic, Z., Charles, F. *et al.*, CMS Note, 1998/006 (1998).
10. Kolb, E. and Turner, M., *The Early Universe*, Palo Alto, CA, Addison Wesley Publishing Company (1990).

Further reading

Bailin, D. and A. Love, *Supersymmetric Gauge Field Theory and String Theory*, Bristol, Institute of Physics Publishing (1994).
Riotto, A. and M. Trodden, Recent progress in baryogenesis, *Ann. Rev. Nucl. Part. Sci.* (1999).
Ross, G., *Grand Unified Theories*, Menlo Park, CA, Benjamin/Cummings Publishing Co. (1985).
West, P., *Introduction to Supersymmetry and Supergravity*, Singapore, World Scientific (1990).

Appendix A
The Standard Model

Science cannot solve the ultimate mystery of nature. And that is because, in the last analysis, we ourselves are part of nature and therefore part of the mystery that we are trying to solve.

Max Planck

There ain't no answer, there ain't going to be any answer. There never has been an answer. That's the answer.

Gertrude Stein

We have put some of the calculational details for the SM in this appendix. For a dimensionless action, S, the Lagrangian, L, and Lagrangian density, ℓ, are defined to be $S = \int L dt = \int \ell d^4 x$, $L = \int \ell d\vec{x}$. The dimension of the density is then $[\ell] = M^4$ since $[S] = 1$. The dimensions of the scalar field are those of mass, $[\phi] = M$, because the "kinetic term" in the Lagrangian density is $\partial \bar{\phi}^* \partial \phi$ and $[\ell] = M^4$. For example, a coupling g to a "potential" term quadratic in the field is dimensionless, $\ell \sim g\phi^4$ $[g] = 1$, which applies to the Higgs potential.

We begin with the SM couplings of fermions to gauge bosons by examining the free particle Dirac Equation. The free particle Lagrangian density, ℓ, for a fermion with wave function ψ, described by the Dirac Equation, with Dirac matrices γ, can be used to find the interaction of the fermion with the photon field, ℓ_I, by making the gauge replacement, $\partial \to D = \partial - ieA$, for the derivative that contains the field A and the charge e. We will use ψ for the fermion fields, ϕ for the scalar fields, and φ for the vector gauge fields. For masses, m is used for fermions, M for bosons.

This replacement should already be familiar, as it appears both in classical mechanics and in non-relativistic quantum mechanics.

$$\ell = \overline{\psi}(i\partial\!\!\!/ - m)\psi, \; \partial\!\!\!/ = \partial_\mu \gamma^\mu. \tag{A.1}$$

The gauge replacement leads to an interaction term in the Lagrangian that has universal coupling of the fermion current, J_μ, to the gauge field, A_μ, with a strength e. Thus, the gauge replacement specifies electrodynamics:

$$\ell_I = e\overline{\psi}\gamma_\mu \psi A^\mu$$
$$= J_\mu A^\mu. \tag{A.2}$$

We proceed by analogy to explore the other forces in the SM. Strong interactions are assumed to be mediated by massless "gluons" universally coupled to the "color charge" of quarks, which are arbitrarily called red, green, and blue (R, G, B), with a coupling constant g_s. Roughly speaking the strong fine structure constant is $\alpha_s = g_s^2/4\pi \hbar c \sim 0.1$, which is ~ 14 times larger than the electromagnetic fine structure constant $\alpha \sim 1/137$. The coupling is not really constant with mass due to quantum loop corrections. In addition the strong coupling constant is only well defined for distances smaller than about 1 fm, where it is less than 1, indicating weak coupling. This means we cannot define the coupling at large distances as we can for electromagnetism. The converse is that the coupling becomes weak at short distances. Therefore, in reactions with high transverse momentum, or short distances, with which we concern ourselves exclusively in this text, we can treat the strong interactions perturbatively.

The labels for the color quantum number, (R, G, B), have no intrinsic meaning. In the interest of brevity we cannot explore in any detail the reasons why we believe there are three colors for quarks. Suffice it to say that the observed strongly interacting particles, such as protons, are colorless because color is "confined" by the strong force that becomes strong at large distances. Therefore, free quarks cannot be observed. In addition, a particle like the uuu bound state (the nucleon resonance Δ^{++}, $J = 3/2\hbar$, $L = 0$) must be overall antisymmetric under exchange since it is a fermion, while it is clearly flavor (uuu), space $(L = 0)$, and spin $(\uparrow\uparrow\uparrow)$ symmetric. The arrows represent the u quark spin directions in the symmetric $J = 3/2$ spin state. An additional degree of freedom, color, must exist and the state must be antisymmetric in color, if uuu is to represent a fermion. That there are three colors comes from comparing the cross section for electron–positron annihilations to muons and quark pairs. In the case of the quarks all color pairs in the final state must be summed over, yielding three times the cross section expected in the absence of color. This factor of three is experimentally confirmed.

The covariant derivative of the fermions (colored quarks) to the vector fields requires the existence of the vector gauge field itself and specifies the universal interaction just as it did for electrodynamics. We assert that the special unitary group in N dimensions, SU(N), has $N^2 - 1$ generators. Thus, the color SU(3), (3 for R, G, B), group has eight colored gluons as its generators. The student need not be conversant with group theory to understand the majority of the material that follows. The gluons can be thought of as $R\overline{B}$, $R\overline{G}$, etc.

The triple vertex of a quark pair and a gluon preserves the color "charge." For example, an $R\overline{G}$ gluon could be emitted by an R quark that then turns into a G quark, as illustrated schematically in Fig. A.1.

The eight massless gluons, g_c, $c = 1, \ldots, 8$, couple to the color triplet (R, G, B) quarks with a universal coupling g_s up to constants that are specified by the SU(3) group properties. We will not explore the SU(3) group constants in this text, as they are not required. The interested student should consult the references at the end of Chapter 1 for more advanced reading.

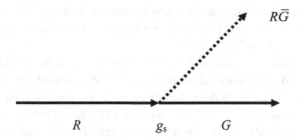

Figure A.1 Schematic representation of a quark–gluon vertex where a red quark emits a red-antigreen gluon and changes into a green quark.

The strong force is therefore developed in very close analogy to the electromagnetic force:

$$U(1) \rightarrow SU(3),$$
$$-ie \rightarrow g_s,$$
$$A \rightarrow g_c. \tag{A.3}$$

Wait just a moment, you may say. The strong interactions are hypothesized to be mediated by massless gluons. Therefore, just as with gravity and electromagnetism, we expect the force to be long ranged, with forces going as the inverse of the square of the distance. However, we know that the nuclear force is very short ranged.

It is far beyond the scope of this text to explore the complete theory of the strong force, quantum chromo-dynamics (QCD). Suffice it to say that this paradox is resolved by realizing that colored objects like quarks are required to be confined to spatial regions defined by the QCD "cutoff" parameter $\Lambda_{QCD} \sim 0.2$ GeV or ~ 1 fm. At this distance and larger, the strong force becomes very strong. This great strength leads to permanent confinement of quarks inside colorless hadrons (like protons, neutrons) and makes the observed force between colorless quark bound states effectively short ranged even though the gluons are massless.

Let us now turn to the weak force. The first theory of weak interactions was proposed by Fermi in the 1930s. It concerned itself with four fermions interacting at a single point with a strength defined by an effective coupling constant G. This theory is not "renormalizable," by which we mean that calculations of higher order processes result in infinities, indicating profound difficulties with the theory. A more fundamental theory was needed and it evolved in the late 1960s and early 1970s. In this theory a close analogy was again made to the successful theory of electrodynamics, which is the prototype of a successful, renormalizable, quantum field theory.

The weak flavor group, with quark and lepton doublets as basic representations, is asserted to be SU(2). Therefore, it has three W boson generators, while the U(1) group of electromagnetism has a single force carrier, the B^o. Weak interactions are mediated by vector bosons, $\mathbf{W} = (W^+, W^o, W^-)$, universally coupled to the weak doublets of quarks and leptons, via weak "charge," or flavor. The electric charge, Q_e, is related to the weak isospin, I_W, projection of the quark or lepton and the "hypercharge," Y_W, which is put in "by hand," $Q = (I_3 + Y/2)_W$. Hypercharge is therefore defined to be $Y_W = -1$ for the leptons and $I_W = 1/3$ for the quarks.

The U(1) group has one generator, B^o, with coupling g_1, while the SU(2) group has three generators $\mathbf{W}$ with universal coupling g_2. The $\mathbf{W}$ is a weak isotriplet, so that it clearly carries weak charge. In this appendix we adopt the simplified, but conventional, notation of W for the field φ_W, Z for φ_Z, and A for the photon field φ_γ. The covariant derivative is constructed to be a scalar in the group space because it appears in the scalar Lagrangian. The covariant derivative of the combined SU(2) $\otimes$ U(1) theory is:

$$D = \partial - i\big[g_1(Y_W/2)B^o + g_2\vec{I}_W \cdot \vec{W}\big]. \tag{A.4}$$

The combined SU(2) and U(1) theory contains two neutral bosons. The Weinberg electroweak mixing angle, θ_W, exists because the physical vector bosons act on the weak eigenstates and not the strong eigenstates (quarks). Thus the two neutral gauge bosons can quantum mechanically mix. We need to write the covariant derivative in terms of the observable electroweak eigenstates, called A and Z:

$$\begin{pmatrix} A \\ Z \end{pmatrix} = \begin{pmatrix} \cos\theta_W & \sin\theta_W \\ -\sin\theta_W & \cos\theta_W \end{pmatrix} \begin{pmatrix} B^o \\ W^o \end{pmatrix},$$

$$D = \partial - i\begin{bmatrix} (g_1 Y/2\cos\theta_W + g_2 I_3 \sin\theta_W)A + g_2(I^+W^- + I^-W^+) \\ + (g_2 I_3 \cos\theta_W - g_1 Y/2\sin\theta_W)Z \end{bmatrix}. \tag{A.5}$$

The coupling to charge is then fixed to be Qe because that is known for the photon.

$$g_1(Q - I_3)\cos\theta_W + g_2 I_3 \sin\theta_W = Qe,$$

$$g_1 \cos\theta_W = g_2 \sin\theta_W = e,$$

$$g_1 g_2 = e\sqrt{g_1^2 + g_2^2}. \tag{A.6}$$

We now see that there is a unification of the weak and electromagnetic force into the "electroweak" force. The charge e is required to be related to the SU(2) coupling as $e = g_2 \sin\theta_W$. The parameter θ_W has been measured and turns out to be a number of order 1. Therefore, at this fundamental Lagrangian level the electromagnetic coupling e has a strength comparable with the "weak" coupling strength. The weak interactions are not intrinsically weak.

Having identified A with the physical photon field and having fixed the photon coupling to be the charge, Qe, we gather up the remaining terms in the covariant derivative that contain the new W and Z bosons.

$$D = \partial - i[eQA + g_2(I^+W^- + I^-W^+) + CZ],$$

$$C = -g_1(Q - I_3)\sin\theta_W + g_2 I_3 \cos\theta_W$$

$$= \sqrt{g_1^2 + g_2^2}(-Q\sin^2\theta_W + I_3\sin^2\theta_W + I_3\cos^2\theta_W)$$

$$= \sqrt{g_1^2 + g_2^2}(I_3 - Q\sin^2\theta_W). \tag{A.7}$$

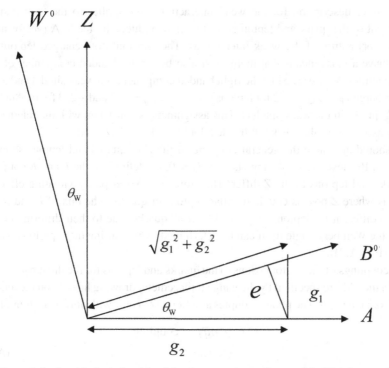

Figure A.2   Graphical relationship of the electroweak couplings and the Weinberg angle.

The W couples to weak isospin, raising and lowering operators, I^+, I^-, so that the W is responsible for the "beta decay" processes where charge changes. The coupling of the Z to fermions is more complicated, and depends on the weak isotopic spin projection, I_3, and the charge, Q, of the fermion. Nevertheless, the Z coupling strength to quarks and leptons is also of order e. In what follows we will replace the notation g_2 by g_W to indicate the weak, SU(2), coupling constant. This notation is also conventional.

$$g_2 = g_W,$$

$$D = \partial - i\left[eQA + g_2(I^+W^- + I^-W^+) + \sqrt{g_1^2 + g_2^2}(I_3 - Q\sin^2\theta_W)Z\right]$$

$$= \partial - i\left[eQA + g_W(I^+W^- + I^-W^+) + g_W/\cos\theta_W(I_3 - Q\sin^2\theta_W)Z\right]. \quad (A.8)$$

The relation of the coupling constants is shown in Fig. A.2. This figure should serve as an aid to memory for the couplings, which are related by the rotation specified by the Weinberg angle.

The Weinberg angle specifies the relationship of the two weak coupling constants, g_1 and g_2, to the electromagnetic coupling constant, e.

$$\cos\theta_W = e/g_1 = g_2/\sqrt{g_1^2 + g_2^2},$$

$$\sin\theta_W = e/g_2 = g_1/\sqrt{g_1^2 + g_2^2},$$

$$\tan\theta_W = g_1/g_2. \quad (A.9)$$

The above description for the weak interaction only applies to the "left handed" component of the quark and lepton wave functions induced by the V-A (vector minus axial vector) nature of the weak interactions. The fact that both charged leptons and quarks have a mass means that there must also be a right handed component of their wave functions. We assume that the right handed component is a weak singlet and assign a hypercharge $Y_R = 2Q_l = -2$ for leptons and $Y_R = 2Q_q = 4/3$ and $-2/3$ for the "up" and "down" quarks in the weak singlets. This assignment is consistent with the relationship already assumed for the weak left handed doublets, $Y_L = 2(Q - I_3)$.

We should also note the general coupling of the Z to quarks and leptons shown in Eq. (A.8). Because the weak singlets have $I_3 = 0$, by definition, the L and R couplings of quarks and leptons to the Z differ. Therefore, we expect parity violating effects in reactions where Z bosons couple to either leptons or quarks. The mix of L and R couplings is different for leptons, "up," and "down" quarks, due to their differing charge. In fact, the Weinberg angle itself can be determined by measuring these parity-violating effects (Eq. (A.7)).

The couplings of the vector bosons to the quarks and leptons for the three basic forces found in the SM are specified by the gauge replacement, drawing heavily on the analogy to electromagnetism. The photon couples as a Lorentz vector with universal strength Qe.

$$\ell^+\ell^-\gamma, q\bar{q}\gamma - \text{coupling}$$

$$Qe\gamma_\mu. \tag{A.10}$$

The weak charge changing coupling of the W to leptons is a Lorentz vector minus axial vector (V-A, parity violating left handed coupling) with a universal strength $g_W = e/\sin\theta_W$.

$$\ell^-\bar{\nu}_\ell W^+ - \text{coupling}$$

$$g_W\gamma_\mu(1 - \gamma_5)$$

$$q\bar{q}'W - \text{coupling}$$

$$g_W\gamma_\mu(1 - \gamma_5)V_{qq'}. \tag{A.11}$$

The coupling to quarks (strong eigenstates) in Eq. (A.11) has an additional vertex factor that is a unitary 3×3 mixing matrix, $V_{qq'}$, which specifies the relationship of the strong and weak eigenstates and preserves the universal weak coupling strength. The mixing matrix of quarks for three generations has, in all generality, two real parameters and one complex number (a total of four parameters) that define it. Therefore, since the matrix elements are not purely real, CP (combined operation of charge conjugation and parity inversion) violation is allowed in the SM for these charge changing weak interactions. In fact, CP violation has been observed in both strange quark and bottom quark decays.

For weak neutral currents the coupling is of similar strength, depends on quark and lepton quantum numbers, and is flavor diagonal (by construction).

$$q\bar{q}Z - \text{coupling}$$

$$g_W(I_3 - Q\sin^2\theta_W)/\cos\theta_W. \tag{A.12}$$

For the strong interactions, the gluons couple as Lorentz vectors to the colored quarks with a universal strength g_s.

$$q\overline{q}g - \text{coupling}$$

$$g_s\gamma_\mu. \tag{A.13}$$

So far we have simply asserted that the weak bosons are massive while the gluons and photons are massless. However, we are not allowed to simply add a vector mass term into the Lagrangian, because it would violate the gauge symmetry. This situation is perhaps familiar from classical electrodynamics. A massive photon cannot preserve the gauge freedom to redefine the electromagnetic potential. Hence, we must indirectly induce a mass for the W and Z.

The weak bosons must have a mass, because the weak interactions are observed to be weak at large distances. The situation is salvaged by introducing a new fundamental scalar field, the Higgs field. The Higgs is chosen to be an electroweak doublet. That is necessary if the boson mass and the fermion mass terms induced by the Higgs field are to be a singlet in the Lagrangian density. It is assumed that the field ϕ possesses a vacuum state where the field is not zero, $\langle\phi\rangle$. The kinetic energy term for this field in the Lagrangian density is $\ell \sim (\partial)\overline{\phi}^*(\partial\phi)$.

For the neutral member of the Higgs doublet, the covariant derivative, Eq. (A.8), with $Q = 0$ is, schematically, $D \sim \partial - ig_W[W + Z/\cos\theta_W]$, involving the W and Z but not the photon. The covariant derivative replacement in the Lagrangian density is:

$$(\partial\overline{\phi})^*(\partial\phi) \to (D\overline{\phi})^*(D\phi).$$

$$\phi \sim \begin{bmatrix} 0 \\ \langle\phi\rangle \end{bmatrix},$$

$$(D\overline{\phi})^*(D\phi) \sim [g_2^2\langle\phi\rangle^2/2]\overline{W}W + [(g_1^2 + g_2^2)\langle\phi\rangle^2/2]\overline{Z}Z + e^2(0)\overline{A}A. \tag{A.14}$$

The result of the gauge replacement in the Higgs Lagrangian is that new quartic terms in the Lagrangian containing the vacuum Higgs field squared are generated. Recall that an explicit mass term for the vector bosons in the Lagrangian density would be of the form $\ell \sim -M^2\varphi^2$, using the relativistic "length" of the momentum vector in the Lagrangian density, $\ell = \overline{\varphi}(P_\mu P^\mu - M^2)\varphi = \overline{\varphi}(\partial_\mu\partial^\mu - M^2)\varphi$ (see also Chapter 1).

Therefore the quartic terms generate specific masses for the W and Z. These masses depend on measurable SM parameters:

$$M_\gamma = 0,$$

$$M_W = g_2\langle\phi\rangle/\sqrt{2},$$

$$M_Z = \langle\phi\rangle\sqrt{g_1^2 + g_2^2}/\sqrt{2} = M_W/\cos\theta_W. \tag{A.15}$$

The numerical values for the masses can now be evaluated. The muon lifetime, $\tau_\mu = 1/\Gamma_\mu$, is determined by the Fermi constant G, $\Gamma_\mu = G^2m_\mu^5/192\pi^3$. In turn, G is an effective coupling, related to the fundamental coupling constant g_W and the boson propagator, which at low momentum transfer is just the boson mass squared.

$$G/\sqrt{2} = g_W^2/8M_W^2, \; G \approx 10^{-5}\text{GeV}^{-2},$$
$$M_W/g_W = \langle\phi\rangle/\sqrt{2},$$
$$\langle\phi\rangle^2 = \sqrt{2}/4G, \; \langle\phi\rangle = 174 \text{ GeV}. \tag{A.16}$$

The W and Z boson masses are induced by the vacuum value of the Higgs field $\langle\phi\rangle = 174$ GeV. The Weinberg angle can be measured in weak "neutral current processes" mediated by the exchange of virtual Z bosons such as $\bar{\nu}_\mu + e^- \rightarrow \bar{\nu}_\mu + e^-$ (see Eq. (A.12)) in addition to the measurements of parity violating effects with lepton and quark pairs in Z mediated interactions. It is important to check that the results of all these measurements give the same result as a test of the consistency of the SM. Using the values of α and θ_w we can find the weak fine structure constant, α_W:

$$\sin^2\theta_W \sim 0.231, \quad \theta_W \sim 28.7°, \quad \sin\theta_W = 0.481;$$
$$\alpha \sim 1/137, \quad \alpha_W = \alpha/\sin^2\theta_W \sim 1/31.6, \quad g_W \sim 0.63. \tag{A.17}$$

Then, from the vacuum value for the field and the weak-coupling constant, the W and Z masses are predicted. These predictions were confirmed in the 1980s with the experimental discovery of both the W and Z particles at CERN in a proton–antiproton collider.

$$M_W = g_W\langle\phi\rangle/\sqrt{2} \sim 80 \text{ GeV},$$
$$M_Z = M_W/\cos\theta_W \sim 91 \text{ GeV}. \tag{A.18}$$

Finally, there are the excitations of the Higgs field about the vacuum state. They are to be interpreted as the field quanta and are, for the Higgs field, labeled as ϕ_H. The interactions of the Higgs field with the bosons of the SM are also fixed by the gauge principle. To see that, we expand the field about the vacuum. In addition to the quartic terms inducing the boson masses there are triplet, $\phi_H WW$ and $\phi_H ZZ$, and quartic, $\phi_H \phi_H WW$, $\phi_H \phi_H ZZ$, interactions of the Higgs quanta ϕ_H with the weak gauge bosons.

$$\phi = \begin{bmatrix} 0 \\ \langle\phi\rangle + \phi_H \end{bmatrix},$$
$$(D\bar{\phi})^*(D\phi) = g_2^2(\langle\phi\rangle + \phi_H)^2\overline{W}W/2 + (g_1^2 + g_2^2)(\langle\phi\rangle + \phi_H)^2\overline{Z}Z/2 \tag{A.19}$$

Clearly the triplet couplings go as $(g_W^2\langle\phi\rangle)\phi_H\overline{W}W + [(g_W^2\langle\phi\rangle)/\cos^2\theta_W]\phi_H\overline{Z}Z$. Since $M_W/g_W \sim \langle\phi\rangle$, these terms are proportional to $g_W M_W$ and $g_W M_Z$ respectively. Thus the Higgs scalar couples to the mass of the gauge vector bosons with weak interaction strength. These terms imply that the Higgs will decay into W and Z pairs, if it is energetically possible. The quartic terms go like $g_W^2\phi_H\phi_H\overline{W}W + [g_W^2/\cos\theta_W^2]\phi_H\phi_H\overline{Z}Z$.

The coupling of the Higgs to fermions is algebraically simple, and is given in Chapter 1 of the text. The mass term identification of $g_f\langle\phi\rangle$ with the fermion mass follows simply from the assumed form of the Yukawa coupling, $g_f\overline{\psi}\phi\psi$, and the Dirac Lagrangian density mass term, $m_f\overline{\psi}\psi$ (Eq. (A.1)). The SM does not specify the fermion couplings to the Higgs, so that no mass prediction is made. However, the hypothesized Yukawa interaction implies that the Higgs quantum couples to fermions with strength proportional to their mass.

Appendix B

A worked example in COMPHEP

Man is a tool using animal . . . without tools he is nothing, with tools he is all.

Thomas Carlyle

Learning is a kind of natural food for the mind.

Cicero

The COMPHEP program is freeware available from its authors at Moscow State University at the site theory.npi.msu.su/~kryukov/comphep.html. There is an online users' manual that is included at the site in addition to the zipped program file that you will download from that site. See also the references at the end of this appendix. Versions for both the Windows® and Linux® operating systems can be downloaded. You are urged to read the users' manual before going further in this appendix.

The COMPHEP program allows us to make Monte Carlo calculations of some sophistication. However, only distributions are calculated and only "tree level" diagrams are included. Thus, we cannot compute individual events using the COMPHEP package alone. In addition, for example, we cannot compute higher order quantum "loops" with this software package. Likewise, decays following production are not directly encompassed in COMPHEP. Finally, the calculations are only made at the fundamental particle level, so that hadronization of the outgoing particles, e.g. quarks and gluons, is not treated in COMPHEP. There are choices for the distribution functions for the initial state proton so proton–(anti)proton reactions can be simulated. Nevertheless, COMPHEP is a complete stand-alone package that we can use to gain considerable insight before attempting to use more complex computer codes.

Help is available using the F1 key. Control is maintained using the Enter, Escape, and Delete keys and the up/down/left/right arrows, as is common in a DOS program.

In the first menu a model is specified. Pick the Standard Model (SM) unless you have a very good reason not to. The next menu has subtasks including "edit model." The lower level tasks are "parameters," "constraints," "particles," and "Lagrangian." The parameters table is shown in Fig. B.1. It is in this table that the Higgs mass is defined, and you can edit it as you wish.

The "constraints" table specifies the W mass in terms of the Z mass and the Weinberg angle (see Appendix A). The remainder of the table defines the CKM matrix, $V_{qq'}$, in terms of the parameters shown in Fig. B.1. The "particles" table is shown in Fig. B.2, and specifies the particles available for COMPHEP calculations. You can edit the SM by changing the "parameters" or "particles" table entries. There are SUSY (see

```
Name  | Value      |> Comment
EE    | 0.31223    |Elementary charge (alpha=1/128.9, on-shell, MZ point, PDG96)
GG    | 1.238      |Strong coupling (LEP/SLD average alphas=0.122, PDG96)
SW    | 0.4730     |sine of the electroweak mixing angle (PDG96)
s12   | 0.221      |Parameter of C-K-M matrix  (PDG96)
s23   | 0.041      |Parameter of C-K-M matrix  (PDG96)
s13   | 0.0035     |Parameter of C-K-M matrix  (PDG96)
Mm    | 0.1057     |muon mass
Mt    | 1.777      |tau-lepton mass        (PDG96)
Mc    | 1.420      |c-quark mass  (pole mass, PDG96)
Ms    | 0.200      |s-quark mass  (pole mass, PDG96)
Mb    | 4.620      |b-quark mass  (pole mass, PDG96)
Mtop  | 175        |t-quark mass  (pole mass)
MZ    | 91.1884    |Z-boson mass           (PDG96)
MH    | 100        |higgs mass
wtop  | 1.7524     |t-quark width           (tree level 1->2x)
wZ    | 2.49444    |Z-boson width           (tree level 1->2x)
wW    | 2.08895    |W-boson width           (tree level 1->2x)
wH    | 0.004244   |Higgs width             (tree level 1->2x)
```

Figure B.1   COMPHEP parameter table for the SM. The first entries specify the three coupling constants at the Z mass. The next three specify elements of the CKM matrix, or the quark mixing matrix. The following masses define the other arbitrary parameters of the SM (see Chapter 6).

```
Full   name  |A   |A+  |2*spin| mass |width  |color
photon       |A   |A   |2     |0     |0      |1
gluon        |G   |G   |2     |0     |0      |8
electron     |e1  |E1  |1     |0     |0      |1
e-neutrino   |n1  |N1  |1     |0     |0      |1
muon         |e2  |E2  |1     |Mm    |0      |1
m-neutrino   |n2  |N2  |1     |0     |0      |1
tau-lepton   |e3  |E3  |1     |Mt    |0      |1
t-neutrino   |n3  |N3  |1     |0     |0      |1
u-quark      |u   |U   |1     |0     |0      |3
d-quark      |d   |D   |1     |0     |0      |3
c-quark      |c   |C   |1     |Mc    |0      |3
s-quark      |s   |S   |1     |Ms    |0      |3
t-quark      |t   |T   |1     |Mtop  |wtop   |3
b-quark      |b   |B   |1     |Mb    |0      |3
Higgs        |H   |H   |0     |MH    |wH     |1
W-boson      |W+  |W-  |2     |MW    |wW     |1
Z-boson      |Z   |Z   |2     |MZ    |wZ     |1
```

Figure B.2   Particles in the SM and their symbolic names. Antiparticles are given, by convention, in upper case. The spins are 0, $\frac{1}{2}$, and 1 and the color representations are singlet, triplet (quarks), and octet (gluons) (see Chapter 1). The neutrinos are defined to be massless, and all stable particles have a zero width assigned to them.

Chapter 6) options in COMPHEP with a much extended particle table, which we do not show here in the interest of brevity. They appear in conjunction with choosing the "MMSM" SUSY model.

The "Lagrangian" shows the explicit Lagrangian that is used in COMPHEP to calculate the matrix elements for all reactions. You can define your own model in the first menu by changing any of the tables discussed so far.

The menu task "enter process" appears next. For this worked example, we choose to study the gluon–gluon production of a *b* quark pair at a CM energy of 100 GeV. The dialogue screen is shown in Fig. B.3.

```
Model:    Standard Model (UnG)

        List of particles (antiparticles)
```

A(A)- photon	G(G)- gluon	e1(E1)- electron
n1(N1)- e-neutrino	e2(E2)- muon	n2(N2)- m-neutrino
e3(E3)- tau-lepton	n3(N3)- t-neutrino	u(U)- u-quark
d(D)- d-quark	c(C)- c-quark	s(S)- s-quark
t(T)- t-quark	b(B)- b-quark	H(H)- Higgs
W+(W-)- W-boson	Z(Z)- Z-boson	

```
Enter  process: G,G -> b,B
```

Figure B.3   Screen capture for the user-entered process of gluon–gluon production of a *b* quark pair. Note the table given with the particle symbolic names for ease of use. A CM energy of 100 GeV is later specified. The option to exclude a set of SM particles from all Feynman diagrams is also available. In this case none are excluded.

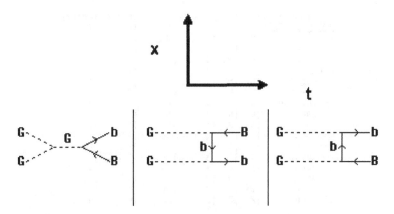

Figure B.4   Feynman diagrams for the process of gluon–gluon production of a *b* quark pair. Time goes left to right and space goes bottom to top by convention. The three diagrams represent gluon–gluon annihilation, and *b* quark exchange.

The next menu has a "view diagrams" subtask. The result for the worked example is shown in Fig. B.4. What is shown is the set of all SM Feynman diagrams for the process the user has specified. There are options in the menu to delete any set of the produced Feynman diagrams. However, we must remember that COMPHEP makes a full quantum mechanically correct complex square of the sums of amplitudes representing the diagrams to get the squared reaction amplitude. Therefore, if any part is excluded, the result for the cross section may not be positive. If you obtain a negative cross section later on you should make sure diagrams are not inadvertently excluded.

After we are satisfied with the diagrams we invoke the menu "squaring" that squares the matrix elements associated with the Feynman diagrams. Then, we invoke the "symbolic

This table applies cuts on the phase space. A phase space function is described in the first column. Its limits are defined and the second and the third columns. If one of these fields is empty then a one-side cut is applied.

The phase space function is defined by a key character and a particle set following this character without separators. For example, "C13" means cosine of angle between the first and the third particles.

The following key characters are available:

```
A  - Angle  in degree units;
C  - Cosine of angle;
J  - Jet cone angle;
E  - Energy of the particle set;
M  - Mass of the particle set;
P  - Cosine in the rest frame of pair;
T  - Transverse momentum P_t  of the particle set;
S  - Squared mass of the particle set;
Y  - Rapidity of particle.
U  - user implemented function.
```
See manual for details.

If you use C-version of this program, you can define the parameter limits by an algebraic formula, which contains numbers and identifiers enumerated in the "Model parameters" menu. Parentheses "()" and operation "+,-,/,*,**,sqrt()" are also permitted.

For the Fortran realization only numbers are permitted into these fields. To define ranges of 'S'-type variable the user must input GeV units value V which will be transformed to V*abs(V).

Figure B.5 Variables available in COMPHEP that can be cut on and whose distributions can be displayed. Options include angle, energy, mass, transverse momentum, or rapidity of a user-specified set of particles.

calculations" menu, which does the spin sum and average appropriate for unpolarized cross sections. We will use COMPHEP as a stand-alone package. Therefore, we do not write out any intermediate results to be used by other Monte Carlo packages. Our aim is to have the student very quickly be able to make a self-contained set of calculations that illuminate the subject matter of the text. Therefore, we only invoke the "numerical interpreter" menu task.

We start by looking at the partonic level for the cross section. To do that we invoke "Vegas" in the next menu. That means performing a Monte Carlo evaluation of the matrix elements and phase space weighting for the quantities in question in order to obtain the cross section. For simple cases the suggested five iterations and 10 000 Monte Carlo trials will go quickly. In other cases the user can appropriately choose the number of trials and number of iterations. Convergence is indicated by a small value of the displayed chi-squared per degree of freedom.

First we do "set distributions." In this example we pick the scattering angle, the angle between the incoming gluon and the outgoing *b* quark. At a fixed CM energy, for two body scattering there is only one free variable and we choose the scattering angle. A list of the available kinematic variables whose distributions can be displayed and that can be cut on is shown in Fig. B.5. Given that a set of several particles can be specified, many different cuts can also be implemented. Particles are labeled sequentially.

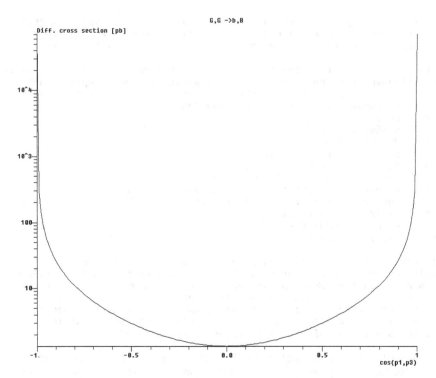

Figure B.6   Angular distribution for the process $g + g \rightarrow b + b$ at CM energy of 100 GeV. Note the forward–backward symmetry due to the fact that the initial state consists of two identical particles. COMPHEP units are pb for cross section and GeV for energy/mass.

In this case the incident gluons are 1 and 2 while the outgoing particles are 3 and 4, as we will see later. You can check the numbering in the process by examining the "sub-process."

In the worked example the chi squared value is 0.66, indicating good convergence. The cross section, integrated over all angles, is 2.65 nb, with a small displayed error on the cross section. The angular distribution is shown in Fig. B.6. It is produced by invoking the "display distributions" task and working through the menus for the number of bins in the histogram, the linear/log choice, and other menu items. The graphical window in COMPHEP is very straightforward, and we leave it to the reader to explore all the options.

Now we know that we are looking at the physics of proton–(anti) proton colliders, and we need to specify how we define the initial state somewhat better. We are in the "Vegas" menu, so hit "Escape" and go to the "IN state" menu. Use escape to back up the menu tree in general. In that menu, select proton on proton at 14 TeV. The dialogue is shown in Fig. B.7.

Setting up for proton–proton collisions at 14 TeV energy in the CM, return to "Vegas" and calculate the cross section using five iterations of 10 000 trials each. The chi-squared

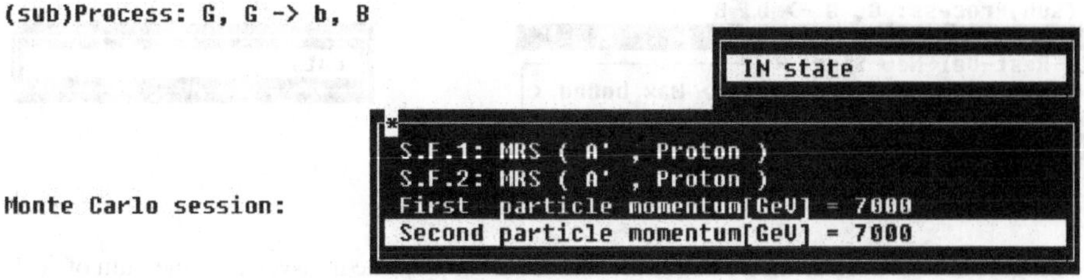

(sub)Process: G, G -> b, B

```
                                    ┌──────────────────────────────┐
                                    │          IN state            │
                                    └──────────────────────────────┘
            ┌───────────────────────────────────────────────────┐
            │█                                                   │
            │ S.F.1: MRS ( A' , Proton )                         │
            │ S.F.2: MRS ( A' , Proton )                         │
Monte Carlo session:  First   particle momentum[GeV]  = 7000     │
            │ Second particle momentum[GeV]  = 7000              │
            └───────────────────────────────────────────────────┘
```

Figure B.7   Selection of protons in the "IN State" dialogue. There are two options for the parameterized distribution functions that are available. Each in turn has two choices of fitted functions. The MRS A data fit is chosen for all calculations in this text, although you may want to try the other, CTEQ, fit in order to convince yourself that the result is insensitive to the choice of distribution functions.

(sub)Process: G, G -> b, B

```
                              ┌───────────────────────────┐
                              │█                          │
                              │ Subprocess                │
                              │ IN state                  │
                              │ Model parameters          │
                              │ QCD scale                 │
                              │ Breit-Wigner              │
Monte Carlo session:          │ Cuts                      │
                              │ Kinematics                │
                              │ Regularization            │
                              │ Vegas                     │
                              │ Simpson                   │
                              └───────────────────────────┘
```

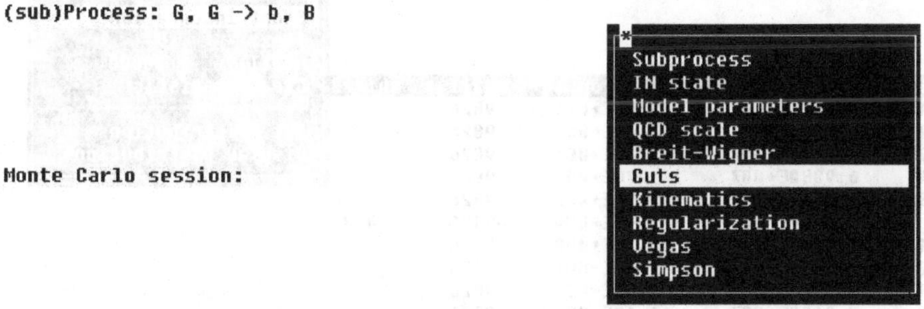

Figure B.8   The menu to set up cuts in the matrix element before doing the phase space integration.

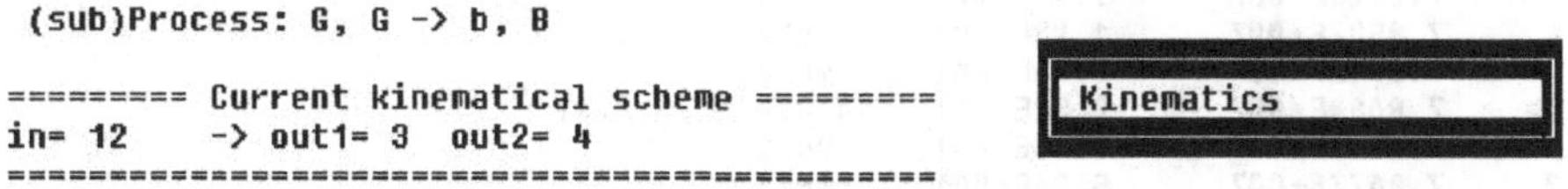

(sub)Process: G, G -> b, B

```
========== Current kinematical scheme ==========          ┌──────────────────┐
in= 12    -> out1= 3  out2= 4                              │   Kinematics     │
================================================           └──────────────────┘
```

Figure B.9   Kinematics labels for particles in the worked example. Particles are numbered sequentially, beginning with the initial state particles.

value is quite large and the cross section is also substantially larger than the partonic cross section we had found. The problem is that the scattering amplitude has a singularity when the scattering angle approaches zero. This is a general feature of "Rutherford" scattering. We avoid it by setting "cuts" in the menu before starting the "Vegas" integration. Possible cuts are explained in Fig. B.8.

The cuts chosen in this particular example are that the transverse momentum of both *b* quarks is greater than 5 GeV. Those cuts exclude arbitrarily small scattering angles because zero angle means zero transverse momentum. The "Kinematics" output is shown in Fig. B.9, while the "cuts" table set by user input is shown in Fig. B.10.

The resulting cut "Vegas" output is shown in Fig. B.11. The value of the chi-square is still very large. That indicates that better-chosen cuts will be needed to obtain a well-behaved solution. The reader should notice that the use of COMPHEP is not just plugging

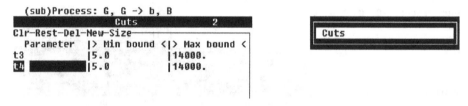

Figure B.10   User defined cuts for the worked example. The transverse momentum of both *b* quarks must be above 5 GeV. The defined cuts are the logical "AND" of the input lines.

(sub)Process: G, G -> b, B

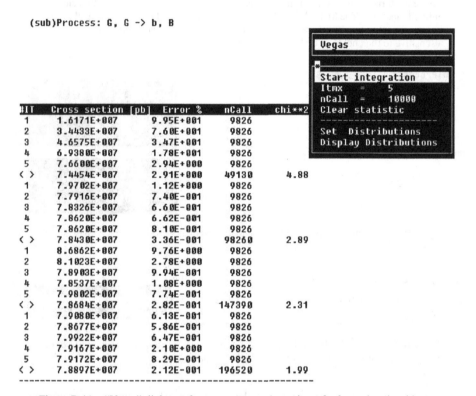

#IT	Cross section [pb]	Error %	nCall	chi**2
1	1.6171E+007	9.95E+001	9826	
2	3.4433E+007	7.60E+001	9826	
3	4.6575E+007	3.47E+001	9826	
4	6.9380E+007	1.78E+001	9826	
5	7.6600E+007	2.94E+000	9826	
< >	7.4454E+007	2.91E+000	49130	4.88
1	7.9702E+007	1.12E+000	9826	
2	7.7916E+007	7.40E-001	9826	
3	7.8326E+007	6.60E-001	9826	
4	7.8620E+007	6.62E-001	9826	
5	7.8620E+007	8.10E-001	9826	
< >	7.8430E+007	3.36E-001	98260	2.89
1	8.6862E+007	9.76E+000	9826	
2	8.1023E+007	2.78E+000	9826	
3	7.8903E+007	9.94E-001	9826	
4	7.8537E+007	1.08E+000	9826	
5	7.9802E+007	7.74E-001	9826	
< >	7.8684E+007	2.82E-001	147390	2.31
1	7.9080E+007	6.13E-001	9826	
2	7.8677E+007	5.86E-001	9826	
3	7.9922E+007	6.47E-001	9826	
4	7.9167E+007	2.10E+000	9826	
5	7.9172E+007	8.29E-001	9826	
< >	7.8897E+007	2.12E-001	196520	1.99

Figure B.11   "Vegas" dialogue for proton–proton creation of a *b* quark pair with cuts made on the *b* quark transverse momentum. Note the large cross section, ~10 microbarns, and the large value of chi-squared.

into a "black box" and waiting for a result. As with most things in life, taste and judgment are called for. As we see in Chapter 4 and Chapter 5, the *b* cross section is, indeed, quite large at the LHC.

The angular distribution for the *b* quark in 14 TeV *p–p* collisions at the LHC when each *b* of the pair has a transverse momentum >5 GeV is shown in Fig. B.12. Note the characteristic Rutherford scattering forward and backward scattering peaks. This feature persists from the gluon–gluon sub process to the overall *p–p* process. Note also that the cross section near 90° scattering is ~10^5 times the *g–g* rate at 100 GeV subenergy.

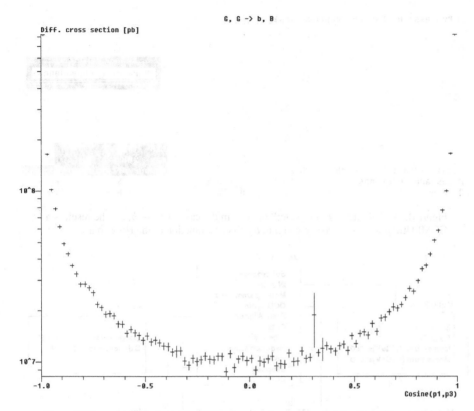

Figure B.12   *p–p* scattering at the LHC, showing the angular distribution of the outgoing *b* quarks in $p + p \rightarrow b + \bar{b}$.

This indicates that much of the cross section arises from gluon scattering at much lower subenergies than 100 GeV, because of the strong energy dependence of the gluon cross section, and the strong x dependence of the gluon structure functions (see Chapter 3).

This completes the worked example. The reader is encouraged to try one or more of his or her own choosing. For processes without free variables, parameter variation and associated graphics are available. The fundamental particle sub-process can be studied in the "Simpson" menu of "Vegas." The "model parameters" menu also lets us change parts of the model. For example, we can vary the Higgs mass. There are useful options for studying decays too. Branching ratios can be found by invoking the "x," or inclusive, particle. For example, entering the process $H \rightarrow 2*x$ gives the decay rates of all two body Higgs decays allowed in the COMPHEP model.

The results can also be written out as .txt files that can then be imported to other programs, and, for example, the results plotted. Indeed, this is the method by which many of the plots shown in the body of the text were made. For example, in $H \rightarrow b + \bar{b}$ there are no free variables, and COMPHEP allows you to vary several parameters. User supplied input varying the Higgs mass is shown in Fig. B.13. The resultant graph of the *b* pair decay width as a function of Higgs mass appears in Chapter 5.

```
Process: H->2*x (5 subprocesses)
```

```
                                                    min= 136
                                                    max= 200
 Total width :        0.004871 GeU                  Number of points= 51
 Modes and fractions :        e2 E2 -    0.02%           s S -      0.17%
 e3 E3 -     4.34%             c C -      8.33%           b B -     87.15%
```

Figure B.13   "Numerical interpreter" options in the case of $H \to b, B$. The result is a COMPHEP graph of the decay width to b pairs as a function of the Higgs mass.

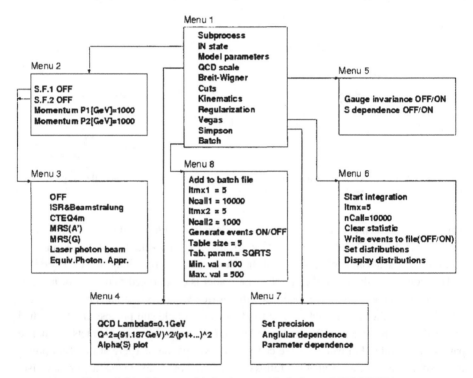

Figure B.14   Menu entries for the numerical phase of COMPHEP.

The Users' Manual appears in the references. This is a comprehensive document. Two figures from that document are shown in Fig. B.14 and Fig. B.15. They show the general flow of the menus in the symbolic and numeric phases of a COMPHEP session.

It is very difficult to fully appreciate the material presented in this text without gaining some facility with COMPHEP, or a comparable program. The reader is strongly

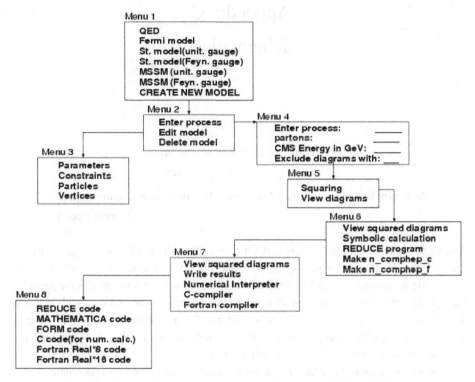

Figure B.15   Menu entries for the symbolic section of COMPHEP.

encouraged to get the most out of this text by gaining a mastery of this program. Some "fiddling around" is very useful to get a feeling for the limits and the power of the COMPHEP program.

Further reading

Kovalenko, A. D. and A. Pukhov, *Nucl. Inst. Meth. Phys. Res. A*, **389** (1997).

Pukhov, A., E. Boos, M. Dubinin, V. Edneral, V. Savrin, S. Schichanin, and P. Semenov, *Archive for COMPHEP User's Manual*, xxx.lanl.gov/format/hep-ph/9908288.

Pukhov, A., E. Boos, M. Dubinin, V. Edneral, V. Savrin, S. Schichanin, and P. Semenov, *User's Manual, COMPHEP V33*, Preprint INP-MSU 98-41/542.

Appendix C

Kinematics

Everything is energy in motion.

Pir Vilayay Khan

In Nature things move violently to their place and calmly in their place.

Francis Bacon

The units we have adopted in this text set $c = 1$. The kinematics of a single particle are specified by the vector momentum, $\vec{P}$, and the rest mass, m, of the particle. The relativistic momentum vector has four components, $P_\mu = (E, \vec{P})$, where E is the particle energy. The relationship between P, E, and m is defined by the velocity with respect to c, $\beta = v/c$, $\gamma = 1/\sqrt{1 - \beta^2}$. The relationships, $E = \gamma m$, $P = \beta \gamma m$, $E^2 = P^2 + m^2$, $P/E = \beta$, can be visualized as a right triangle having sides m and P, with hypotenuse E, or sides and hypotenuse 1, $\beta\gamma$, and γ respectively.

Now we move on to the phase space for a single particle. The non-relativistic phase space, Eq. (C.1), for a single particle is familiar from classical Maxwell–Boltzmann statistics. It states that all Cartesian momentum components are equally probable. The magnitude of the particle momentum is P. The momentum component parallel to the beam is labeled by $P_\parallel$, while the perpendicular component is defined to be P_T. The solid angle element is $d\Omega$ and the azimuthal angle is ϕ.

$$d\vec{P} = dP_x dP_y dP_z = P^2 dP d\Omega = dP_\parallel P_T dP_T d\phi. \tag{C.1}$$

The relativistic generalization of the classical one body phase space is given in Eq. (C.2), where y is a kinematic variable called the rapidity. The one particle phase space is simply the four-dimensional momentum volume with a constraint that the particle has a fixed mass set by the sharply peaked Dirac delta function, δ. The rapidity is the relativistic analogue of longitudinal velocity. Particle energy is E, so as $E \to m$, $dy \to dv_\parallel$ and Eq. (C.1) is recovered in the limit.

$$d^4 P \delta(E^2 - P^2 - m^2) = d\vec{P}/E = P_T dP_T d\phi dy,$$
$$dy = dP_\parallel/E. \tag{C.2}$$

If the transverse momentum is limited by dynamics, we expect a uniform distribution in y for any particle produced in an inelastic collision if the momentum carried off by the produced particle is small. In general, we will see that almost all produced particles are uniformly distributed in rapidity, at least at wide angles (small rapidity) with respect to the beam.

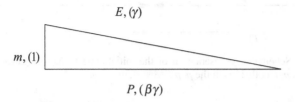

$E, (\gamma)$

$m, (1)$

$P, (\beta\gamma)$

Figure C.1   The relationship of the rest mass, m, the momentum, P, and the energy, E.

We assert that rapidity adds under Lorentz transformation. Thus, rapidity is the relativistic generalization of velocity. Note also that the one particle phase space is uniformly distributed in (y, ϕ) area for small y. The rapidity defined above is approximated by the pseudorapidity variable defined in Chapter 2 if the particle masses are small with respect to the transverse momentum. Therefore, the detector shown in Chapter 2 is segmented into "pixels" of equal one particle phase space by design. This fact also serves as a belated justification of the use of (η, ϕ) coordinates in the plots shown in Chapter 2 and later in the body of the text.

We can integrate the expression given in Eq. (C.2), $dy = dP_{\parallel}/\sqrt{P_{\parallel}^2 + P_T^2 + m^2}$, to find the relationship between energy and rapidity.

$$E = m_T \cosh y,$$
$$m_T^2 = m^2 + P_T^2. \tag{C.3}$$

We can also derive this formula using the relationship between E, P, and m. The identity is $E^2 - P_{\parallel}^2 = P_T^2 + m^2 \equiv m_T^2$. Comparing that with the hyperbolic identity, $\cosh^2 y - \sinh^2 y = 1$, we can easily confirm Eq. (C.3) and, in addition, find that $\sinh y = P_{\parallel}/m_T$, $\tanh y = P_{\parallel}/E$.

Therefore, for massless single particles, or particles with mass much less than transverse momentum, $m_T \sim P_T$, where $P_T = E \sin \theta$:

$$\cosh y = 1/\sin \theta,$$
$$\sinh y = 1/\tan \theta,$$
$$\tanh y = \cos \theta. \tag{C.4}$$

In this particular limiting case, we can find a simple relationship between polar angle and rapidity. Using Eq. (C.4) we can easily show that

$$e^{-y} = \tan(\theta/2). \tag{C.5}$$

Therefore, in this limit we are justified in using the equality of the rapidity, y, and the pseudorapidity, η.

Now let us move from single particle kinematics to the kinematics of two particle systems. We specialize in the case of two partons contained within the proton and (anti)proton defining the initial state. We further assume that the frame we use is the proton–(anti)proton CM frame. As seen in Fig. C.2, the partons have longitudinal

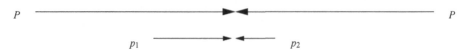

Figure C.2   Schematic representation of the initial state in parton–parton scattering
starting from p–p collisions in the p–p CM system.

momentum $p_1 = x_1 P$ and $p_2 = x_2 P$ respectively, where P is the momentum of the proton
in the p–p CM. The quantity x is the fraction of the proton momentum carried by the
"parton" or the fundamental point like constituent that exists within the proton.

The mass, M, and momentum fraction, x, of the initial state is then found by
conservation of relativistic energy and momentum. The four-dimensional momentum
$P_\mu = (E, \vec{P})$ has an invariant "length" of $P_\mu \cdot P^\mu = M^2$. We simply assert that the single
particle relationships carry over to systems of particles. For example, in the initial p–p
state in the CM, $(P_1 + P_2)_\mu = (E_1 + E_2, \vec{0}) \sim (2P, \vec{0})$. The CM energy squared, s, is
$s = (P_1 + P_2)_\mu \cdot (P_1 + P_2)^\mu = (E_1 + E_2)^2 \sim 4P^2$. The mass of the two parton system,
M, follows, assuming that the partons are massless and have no transverse momentum:

$$M^2 = (p_1 + p_2)_\mu \cdot (p_1 + p_2)^\mu \sim (e_1 + e_2)^2 - (\vec{p}_1 + \vec{p}_2)^2$$
$$\sim P^2[(x_1 + x_2)^2 - (x_1 - x_2)^2],$$
$$x = p_\parallel/P \sim 2p_\parallel/\sqrt{s}. \tag{C.6}$$

A bit more algebra allows us to find M and x for the initial state in terms of x_1 and
x_2, $x_1 x_2 = M^2/s = \tau$, $x_1 - x_2 = x$. A typical value, $\langle x \rangle$, for the momentum fraction
of the parton producing a state of mass M at p–p CM energy $\sqrt{s}$ occurs when $x_1 = x_2$
or when $\langle x \rangle$ is equal to $\sqrt{\tau}$. For example, top quark pairs at the Tevatron, with $M \sim$
$2m_t \sim 350$ GeV, are produced at rest in the CM by partons with momentum fraction
$\langle x \rangle \sim M/\sqrt{s} = 350/1800 \sim 0.2$.

Having produced the initial state, we assume it "decays" into a two body final state.
Schematically, the reaction is $1 + 2 \rightarrow 3 + 4$. In a two body "decay" the transverse
momentum of each massless final state parton is a function of the mass of the decaying
state and the decay angle, $P_{T3} = P_{T4} = E_T = (M/2)\sin\theta$.

The measured values of the two parton kinematic quantities, y_3, y_4, and E_T allow us
to solve for the variables x, M, and $\hat{\theta}$. Using results given above we can relate M and x
to the initial state defined by x_1 and x_2, thus completely specifying the kinematics for
the two body process. These relationships follow from the conservation of energy and
momentum and the definition of rapidity given above. For example, the mass can be
expressed, using $E_T = M_T \cosh y$, $P_\parallel = M_T \sinh y$, in terms of the measured final state
variables:

$$M^2 = 2E_T^2[\cosh(y_3 - y_4) - \cos(\phi_3 - \phi_4)]. \tag{C.7}$$

For "back to back" the limit is $M^2 \rightarrow 2E_T^2[\cosh(y_3 - y_4) + 1]$. If $y_3 = y_4$ then $M \rightarrow$
$2E_T$ or $\hat{\theta} = 90°$.

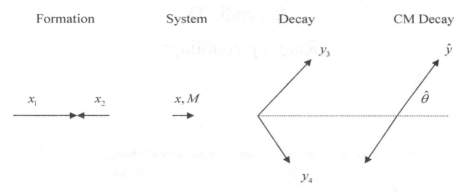

Figure C.3 Schematic representation of two body parton scattering. The initial state partons are found in the proton and (anti)proton. They form an intermediate state of mass M moving with momentum fraction x, rapidity y. This state then "decays" into a two body final state with measured transverse momenta, E_T, and rapidities y_3 and y_4 in the p–p CM frame.

Some kinematic definitions for the two body final state are shown in Fig. C.3. We note that the initial two body state is not the parton–parton center of momentum system in general, although it is, on average. Therefore, the composite state, x and M, is moving in the overall proton–(anti)proton CM system. Thus, in the proton–(anti)proton frame the two body final state is not back to back in polar angle, as it is in the final state CM frame.

Briefly, in the parton–parton CM reference frame, $\hat{y}_3 = \hat{y}$, $\hat{y}_4 = -\hat{y}$ and, since rapidity is additive under Lorentz transformation, in the proton–proton CM, $y_3 = y + \hat{y}$, $y_4 = y - \hat{y}$, where y is the rapidity of the two body state in the p–p CM. Thus the system rapidity y and the parton–parton CM jet rapidity $\hat{y}$ can be found in terms of y_3, y_4, $y = (y_3 + y_4)/2$, $\hat{y} = (y_3 - y_4)/2$.

In the decay, energy and momentum conservation require that each massless parton has an energy/momentum of half the system mass, M, and transverse momentum $P_T = E_T = (M/2)\sin\hat{\theta}$. We can also find the parton–parton CM scattering angle in terms of $\hat{y}$, $\tanh\hat{y} = \cos\hat{\theta}$ (see Eq. (C.4)). Thus with E_T experimentally measured for the partons and $\hat{y}$ found in terms of y_3 and y_4, we can solve for M and $\hat{\theta}$, for example $M = 2E_T/\sin\hat{\theta} = 2E_T/\cosh\hat{y}$. Finally, y and M give us x and M that can be used to solve for the initial state parton momenta, x_1, x_2, by way of $\sinh y = p_\parallel/m_T = xP/M$, $x = (2M/\sqrt{s})\sinh y = M/\sqrt{s}[e^y - e^{-y}] = x_1 - x_2$.

$$x_1 = [M/\sqrt{s}]e^y,$$
$$x_2 = [M/\sqrt{s}]e^{-y}. \tag{C.8}$$

Thus, from measurements of the two body final state we can infer the x values of both initial state partons, and measure the scattering angle.

Appendix D

Running couplings

Everything changes, nothing remains without change.

Buddha

You can run but you can't hide.

Anonymous

In quantum field theory the coupling "constants" of the three SM forces are put into the theory explicitly in the covariant derivative, which enters the basic Lagrangian (Appendix A). These couplings take on "effective" values, which are functions of the mass scale at which they are examined. This effect is due to quantum corrections caused by higher order diagrams.

This effect was first derived in quantum electrodynamics, QED, where it was found that the electron charge increases as we look at small distances. This is understood in physical terms as due to the existence of virtual electron–positron pairs in the vacuum due to the virtual decays of virtual photons emitted and then reabsorbed by the charge. These virtual charged pairs cause charge screening. In a polarizable dielectric medium an induced dipole moment reduces the applied field, which effectively reduces the squared charge by the dielectric constant, ε. Thus, the effect is called "vacuum polarization."

A schematic representation of an electron–positron loop is shown in Fig. D.1.

The electron charge is shielded by virtual γ fluctuations into $e^+ + e^-$ pairs on a distance scale set by the electron Compton wavelength, $\lambda_e \sim \hbar c/m_e \sim 400$ fm. Thus α increases as the mass scale decreases and electromagnetism gets slowly stronger as the mass increases. Conceptually the "bare" charge is surrounded by pairs. One particle of the virtual pair is attracted to the oppositely charged main charge, thus polarizing the vacuum. Therefore, an observer at a given distance from the charge will see the charge reduced, or "shielded," by an amount that decreases with distance.

We assert that the "renormalized" charge at first order in perturbation theory, $e_R(Q^2)$, is given in Eq. (D.1), where m is the electron mass and Q is the mass scale at which the charge is measured.

$$e_R^2(Q^2) \sim e^2[1 + \alpha/12\pi \ln(Q^2/m^2)]. \tag{D.1}$$

The effect is first order in the fine structure constant and depends logarithmically on the mass scale of observation Q. Let us now go ahead with a bit more mathematical detail to see if we can understand that dependence. The schematic representation for the simple case of the charge of a very heavy source, i.e. one suffering no recoil in emitting

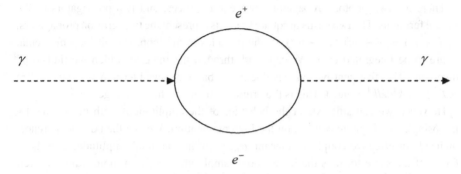

Figure D.1 A photon virtually decays into an electron–positron pair with that pair subsequently annihilating into the original photon.

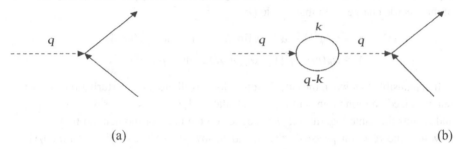

(a)   (b)

Figure D.2 Kinematic definitions for a very heavy source of photons interacting with a fermion of mass m in lowest order (a) and with a virtual electron–positron loop in the next highest order (b).

a photon, is shown in Fig. D.2. The photon virtually decays into a fermion pair which then annihilates into the original photon prior to interacting with the external fermion.

To lowest order, the charge, e, is the "bare" charge that appears in the Lagrangian, and the propagator is the Fourier transform of the Coulomb interaction potential, $V(r)$. In the Born approximation we take the initial $|i\rangle$ and final $|f\rangle$ states to be free particle plane wave states leading to the amplitude A: $A \sim \langle f|V(r)|i\rangle \sim \int e^{i\vec{k}_f \cdot \vec{r}} V(r) e^{-i\vec{k}_i \cdot \vec{r}} d\vec{r} \sim \int e^{i\vec{q} \cdot \vec{r}} V(r) d\vec{r} = V(\vec{q})$. The momentum transfer q is $q^2 = |\vec{q}|^2$, $\vec{q} = \vec{k}_f - \vec{k}_i$. In the case of electromagnetism, $V(r) \sim \alpha/r$, $V(q) \sim \alpha/q^2$. Thus the reaction amplitude in lowest order is $A_o \sim \alpha/q^2$ (Rutherford amplitude).

The loop integral is indicated in Eq. (D.2) and it can be roughly read off by examining Fig. D.2. The factors of γ are Dirac matrices that refer to the vector nature of the interaction vertex (see Appendix A) and the "slash" notation is defined to be $\displaystyle{\not{a}} = a_\mu \gamma^\mu$. Knowledge of the Dirac matrices is not needed to roughly understand the argument, however.

The change in the amplitude due to the existence of the higher order process shown in Fig. (D.2b) is:

$$\delta A \sim \alpha^2 \int dk^4 (1/q^2)[1/(\not{k} + m)][1/((\not{q} - \not{k}) + m)](1/q^2). \tag{D.2}$$

There are two photon propagators, two extra vertices, and two propagators of the virtual fermions. The two terms in square brackets represent the two fermion propagators, $1/(\not{k} + m) = (\not{k} + m)/(k^2 - m^2)$, of the particles in the loop, and all loop momenta, k, are to be integrated over. We can work through the integral, which we find to be divergent. The divergent behavior is due to the badly behaved term, $\delta A \sim \int dk^4 (m/k^2)$ $(m/k^2) \sim \int k^3 dk/k^4 \sim \ln(k)$. This is the origin of the logarithmic divergence.

However, we can still extract the behavior of the amplitude A with mass scale by imposing a cutoff parameter, Λ, on the loop momentum. We can then define a "renormalized" or effective coupling constant, α_R, such that the total amplitude, $A \sim A_0 + \delta A$, is of the same form as the lowest order amplitude at a given momentum transfer, $A \equiv \alpha_R(q^2)/q^2$, with $\alpha_R(q^2) = \alpha_R(m^2)[1 + \alpha_R(m^2)/12\pi \ln(-q^2/m^2)]$. Comparing with Eq. (D.1), we see that we have reproduced the lowest order expression for the behavior of the electric charge, with mass scale $Q^2 = -q^2$.

$$A \sim \alpha/q^2\{1 - \alpha/12\pi\,[\ln(\Lambda^2/m^2) + \ln(-q^2/m^2)]\},$$
$$A \sim \alpha_R(m)^2/q^2[1 + \alpha_R(m^2)/12\pi \ln(-q^2/m^2)]. \tag{D.3}$$

It is plausible that when the calculation is done to all orders in perturbation theory the renormalized coupling constant is also calculable $(1/(1-x) = 1 + x + x^2 + x^3 + \cdots)$ and retains the same logarithmic dependence on the mass or momentum transfer scale at which the reaction proceeds that it did in lowest order, $\alpha_R(Q^2) = \alpha_R(m^2)/[1 - (\alpha_R(m^2)/12\pi)\ln(Q^2/m^2)]$.

It is most natural to see how the inverse of the fine structure constant "evolves," as quoted in Eq. (D.4). The difference in the inverse of the renormalized fine structure constant depends logarithmically on the ratio of the squares of the masses, Q and m, at which they are observed.

$$1/\alpha_R(Q^2) = 1/\alpha_R(m^2) - 1/12\pi\ln(Q^2/m^2). \tag{D.4}$$

For electromagnetism, we can take the charge to large distances as a way to operationally define α. Conventionally, the fine structure constant is defined at large distances, or low masses, to be $\alpha = \alpha(0) \sim 1/137$. Experimentally, at the Z mass, $\alpha(M_Z) = 1/129$. The coupling only becomes strong, $1/\alpha(\Lambda_{QED}^2) = 0$, at an enormous energy, $\Lambda_{QED} \sim m_e e^{(6\pi/\alpha)}$ (see Chapter 6 on the GUT scale). Thus, the running coupling constant scheme can be used for all mass scales of practical interest.

In QCD a similar effect occurs, but with the added complication that the gluons mutually interact whereas the photons are uncharged. The mutual self-coupling of gluons leads to the result that the strong coupling strength actually decreases as the mass increases, opposite to the behavior of electromagnetic charge. The anti-screening of the colored gluons overcomes the screening effects of the colored quarks. As seen in Appendix A (Fig. A.1) the virtual emission of colored gluons will remove color from the vicinity of the "source" quark, and that results in color anti-screening. The "running" of the coupling constant in QCD means that as $Q^2 \to \infty$, $\alpha_s(Q^2) \to 0$.

$$1/\alpha_s(Q^2) = 1/\alpha_s(m^2) + [(33 - 2n_f)/12\pi]\ln(Q^2/m^2). \tag{D.5}$$

In Eq. (D.5), n_f is the number of fermion generations that are "active," or above threshold to occur in the quantum loops at the mass scale Q in question. The fermion term is negative (screening) with a magnitude familiar from QED (Eq. (D.4)). The gluons appear as the positive factor 33 indicating that they anti-screen the color charge. Clearly the gluon effect dominates and the overall effect is anti-screening.

This has profound implications for quarks. As the distance increases, the force gets stronger, ultimately causing permanent confinement of quarks within the hadrons, such as protons, which are themselves colorless. Conversely, the strong interaction becomes weak at high mass scales. Indeed, that is why we focus on high transverse momentum phenomena in this text. The strong interactions are simple and perturbatively calculable in this region of phase space.

For the strong interactions, we cannot separate the charges since the coupling is strong at large distances (low energies). Instead, using Eq. (D.5), we define an energy Λ_{QCD} where the interactions become strong, $\alpha_s(\Lambda_{QCD}^2) \sim \infty$, $1/\alpha_s(\Lambda_{QCD}^2) = 0$, $\Lambda_{QCD} \sim 0.2$ GeV. The strong coupling is observed to run (Chapter 4) and is now conventionally defined at the Z mass, $\alpha_s(M_Z^2) \sim 0.13$. Thus the strong fine structure constant is well defined for mass scales >0.2 GeV.

$$\alpha_s(Q^2) = 12\pi/[(33 - 2n_f)\ln(Q^2/\Lambda_{QCD}^2)]. \tag{D.6}$$

The situation for the weak interactions is analogous to the strong interactions. The weak bosons are themselves carriers of the electroweak charge, and they anti-screen. The fermions screen, but the net effect is again anti-screening. The result is that

$$1/\alpha_W(Q^2) = 1/\alpha_W(m^2) + [(22 - 2n_f - 1/2)/12\pi]\ln(Q^2/m^2). \tag{D.7}$$

The factor of 22 comes from anti-screening of the weakly "charged" W and Z, while the fermion term is now familiar. The new term of $-1/2$ is due to the existence of Higgs in the electroweak loop.

These three coupling constants are used in Chapter 6, along with their supersymmetric generalizations. In addition, we quote the evolution of the W and Z mass due to quantum loops in Chapter 4, the running of the strong coupling in Chapter 4, and the evolution of the Higgs mass with mass scale in Chapter 5. Clearly, the "running" of constants appearing in the Lagrangian is a basic effect of quantum field theory. It is also now part of the precision measurements available in high energy physics and a way to indirectly probe very high mass scales.

An example in the Grand Unified Theories is the running of the masses with the scale. The mass of a state can be defined by the behavior of the propagator. For example, a massive boson has a propagator $1/(q^2 + M^2)$. However, the propagator is modified by quantum loops and thus the mass itself runs. Assuming the SU(5) relation that the τ lepton and b quark have equal mass at the GUT scale, the mass ratio evolved to a lower scale Q is:

$$[m_b(Q)/m_\tau(Q)] = [\alpha_3(Q)/\alpha_{GUT}]^{1/(4\pi^2 b_3)}[\alpha_1(Q)/\alpha_{GUT}]^{-1/(16\pi^2 b_1)}. \tag{D.8}$$

The terms b_1 and b_3 are defined in Section 6.4.

This relationship predicts fairly well the observed mass ratio at approximate GeV mass scales. The student is encouraged to plot Eq. (D.8) and examine the running behavior of the masses. Note that the weak interaction does not contribute to Eq. (D.8) because the Dirac mass term, or the self-energy Feynman diagram, connects left and right handed Dirac spinors and the weak interaction is solely left handed, by construction. Note also that the general reason that quarks are predicted in SU(5) to be heavier than charged leptons at GeV mass scales is that quarks have strong interactions and the strong interactions are strong at low mass scales.

Index

Page numbers in italic refer to figures. Page numbers in bold denote entries in tables.